The New Solar Physics

AAAS Selected Symposia Series

Published by Westview Press
5500 Central Avenue, Boulder, Colorado

for the

American Association for the Advancement of Science
1776 Massachusetts Ave., N.W., Washington, D.C.

The New Solar Physics

Edited by John A. Eddy

AAAS Selected Symposium **17**

AAAS Selected Symposia Series

Published in 1978 in the United States of America by
    Westview Press, Inc.
    5500 Central Avenue
    Boulder, Colorado 80301
    Frederick A. Praeger, Publisher

Library of Congress Catalog Card Number: 78-66338
ISBN: 0-89158-444-7

Printed and bound in the United States of America

About the Book

The new astronomy of the late nineteenth century was built
on spectroscopy; with this tool, Huggins, Lockyer, and Hale
started solar physics on its present course of increasingly
finer examination of increasingly smaller bits of the solar
surface, with ever more sophisticated theoretical analysis.
But after a hundred years of study, we are now finding that
there may be something to learn from a very different approach.

In the last few years, our concepts of the sun have been
altered by four new developments--the discovery of apparent
global solar oscillations, an unsettled and unsettling deficit of neutrinos from the center of the sun, a new elucidation of the role of solar wind, and some disturbing historical
facts that shake old concepts of solar constancy and regularity. Though these developments are probably unrelated,
they all come from an unconventional direction--from outside
the traditional areas of observational and theoretical solar
physics. Significantly, each deals with the sun as a whole
object. This volume brings together summaries of these four
developments in solar physics, written by the four scientists
whose work has prompted our new assessment of the sun.

About the Series

The *AAAS Selected Symposia Series* was begun in 1977 to provide a means for more permanently recording and more widely disseminating some of the valuable material which is discussed at the AAAS Annual National Meetings. The volumes in this *Series* are based on symposia held at the Meetings which address topics of current and continuing significance, both within and among the sciences, and in the areas in which science and technology impact on public policy. The *Series* format is designed to provide for rapid dissemination of information, so the papers are not typeset but are reproduced directly from the camera-copy submitted by the authors, without copy editing. The papers are organized and edited by the symposium arrangers who then become the editors of the various volumes. Most papers published in this *Series* are original contributions which have not been previously published, although in some cases additional papers from other sources have been added by an editor to provide a more comprehensive view of a particular topic. Symposia may be reports of new research or reviews of established work, particularly work of an interdisciplinary nature, since the AAAS Annual Meetings typically embrace the full range of the sciences and their societal implications.

WILLIAM D. CAREY
Executive Officer
American Association for
the Advancement of Science

Contents

List of Figures

Chapter 4

List of Tables

About the Editor and Authors

John A. Eddy is a senior scientist at the High Altitude Observatory, National Center for Atmospheric Research, specializing in solar physics, archaeoastronomy, and the history of astronomy. He has published several articles on solar physics in journals such as Science and Scientific American, and his work was the subject of a 1977 NOVA television program, "The Sunspot Mystery."

Raymond Davis, Jr., senior scientist at the Brookhaven National Laboratory, specializes in nuclear chemistry and cosmochemistry. He has published several articles on solar neutrino detection and was awarded the Boris-Pregel Prize by the New York Academy of Sciences.

John C. Evans, Jr., senior research scientist at the Battelle Northwest Laboratory in Richland, Washington, holds a Ph.D. in chemistry from the University of California at San Diego. His research has focused on cosmochemistry, cosmic rays, solar neutrinos, radiochemistry, and general analytical chemistry.

Henry A. Hill is a professor in the Department of Physics at the University of Arizona. He has conducted experimental tests of general relativity in addition to doing research in solar physics.

Arthur J. Hundhausen, senior scientist at the High Altitude Observatory, National Center for Atmospheric Research, Boulder, is conducting research in solar and terrestrial physics. He has studied extensively the physics of interplanetary space, and his numerous publications include Coronal Expansion and Solar Wind (Springer-Verlag, 1972).

E.N. Parker is a professor in the Departments of Physics and Astronomy and Astrophysics at the University of Chicago. His main area of interest is the classical physics of hydrodynamics and magnetic fields, and he is the author of Interplanetary Dynamical Processes *(Wiley/Interscience, 1963).*

Preface

The five sections of this volume have grown from a symposium of the same name held in February of 1977 at the Annual Meeting of the American Association for the Advancement of Science in Denver. Only advertising agents or other enthusiasts dare label anything "new" in science before time has proven it so, but it seemed to us that there were recent developments in the study of the sun that together asked for a reassessment of past views, and perhaps, the definition of a new kind of solar physics. These included the apparent discovery of global solar oscillations, an unsettled and unsettling deficit of neutrinos from the interior of the sun, a new elucidation of the role of the solar wind and its connection to coronal holes, and some disturbing facts from history that shake old concepts of solar regularity. They may be unrelated. But they have in common that each comes from a nonconventional direction -- from outside the traditional areas of observational and theoretical solar physics. Each deals, perhaps significantly, with the sun as a whole object. The New Astronomy of the late 19th century was built upon spectroscopy and with this tool Huggins, Lockyer, and Hale started solar physics on its present course of finer and finer examination of smaller and smaller bits of the solar surface, with ever more sophisticated theoretical analyses. As Professor Parker explains in his introduction, this course has brought many benefits to all of astronomy. But after a hundred years of close examination of the bark, we find that there may be something important to learn from looks at the tree itself, and perhaps at the forest.

At the Denver symposium, as in this volume, we were able to bring together some of the scientists directly involved in these new ways of looking at the sun. In addition, Eugene Parker of the University of Chicago, who in his own distinguished work has known the feel of new ideas and the force of their results, has given us an introduction that sets solar physics, old and new, in broad and proper perspective.

Each of the ensuing sections tells an original story in its own way, for in editing I have not attempted to force them into a single mold or single style. Some new developments, including my own summary of the derivation of solar history from tree-rings, are more naturally told in general terms. Others, such as Dr. Hill's discussion of the evidence and implications of solar oscillations, need at this time to be explained in depth, with more thorough physical and mathematical development, and we have been able to devote a major part of the volume to an extensive review of this new and important question. In equally important sections on solar neutrinos, by Raymond Davis and John Evans, and on the solar wind by Arthur Hundhausen, we have also captured authoritative and timely reviews. The important thing, it seemed to me, was to keep these new developments in the words of the physicists involved, who were, happily, our authors.

Dr. Hundhausen's chapter, "Streams, Sectors, and Solar Magnetism" originally appeared in the 1977 monograph report of the first Skylab Solar Workshop, Coronal Holes and High Speed Wind Streams, edited by Jack B. Zirker; it is reproduced here, with minor changes, with the kind permission of the editor and the publisher, the Colorado Associated University Press.

Clearly absent is a final chapter that draws these disparate developments together, explaining their connections and their combined significance in our understanding of the sun and other stars. It is far too early for that, and probably unwise to go beyond the obvious conclusion that the global sun is changing and not yet fully understood. There is danger in assuming that the most recent developments in any field are related, simply because they were concurrently found. Yet I think these solar developments will prove connected, or rather tangled together with other roots that we have yet to find, for everything pulled from nature seems connected to everything else.

I am indebted to each of the authors who contributed their time and new ideas to the volume, to Kathryn Wolff of the AAAS Publications Division for her extended patience, and to Mrs. Barbara Kirwin for preparing the typescript.

John A. Eddy
April 1978

The New Solar Physics

Solar Physics in Broad Perspective

E. N. Parker

Past studies of the sun have produced the basic theoretical foundation blocks of modern astrophysics. With recent developments in UV and X-ray observations, high resolution Doppler and magnetic observations, neutrino observation, and the direct observation of the gaseous and particle emission from the sun, together with the associated theoretical problems and developments, it looks as though solar physics will continue as the principal cutting edge of the hard science that can be accomplished with astrophysical research.

Solar physics does not occupy the center of the world of fashion, of course. The philosophical excitement of the open or closed universe question, and the variety of exotic objects discovered in the last two decades, have fired the enthusiasm and occupied the attention of the majority of astrophysicists. The period has been the most productive of new objects and new problems in the history of astronomy. The quasar, the pulsar, the Seyfert galaxy, the intense infra-red galaxy, the galactic and extragalactic X-ray sources (hinting at black holes and accretion disks) and the universal black body radiation are among the principal box office attractions of the glittering astrophysical universe.

None of the active objects now known in the sky could have been anticipated. Indeed, some are still without basic theoretical explanation after years of study, and none are understood beyond ideas as to the general possibilities. The concept of the neutron star, pointed out by Oppenheimer forty years ago, the Schwarzschild singularity, now called the black hole, and the ideas of Gamow on the residual black body radiation from the original compact state of the universe, existed in previous years. The new observational discoveries awakened these slumbering ideas and gave them real form.

Ingenuity in the development of new instrumentation, permitting new observations, and ingenuity in the interpretation of old observations, are responsible for the wealth of discovery. The many new active and variable objects have opened our minds to the enormous variety of effects to be found in the universe beyond the classical concept of the static star and the dynamics of a cloud of stars. The discoveries have indicated the importance of new fields of theoretical physics, such as relativistic plasmas, quantum electrodynamics in strong magnetic fields, and pair creation and partical interactions in the enormous gravitational fields surrounding black holes of small mass. The discoveries have provided the motivation for the rapid development of these subjects, previously too far removed from experience to merit the attention of the theoretician.

We should capitalize on the present era of discovery, while it lasts, but at the same time we should be aware that there are severe limits to how far we can go in the study of fascinating, but distant, objects. The magnitude of the distant active phenomenon can be determined, and the probable or possible nature of the mechanism speculated. But the mechanism invariably involves effects over dimensions far too small to be observed from a distance. Except for very long baseline interferometry and the Michelson interferometer, even the nearest stars are only points of light. The common belief today, that they are glowing spheres of gas like the sun, is entirely a matter of theoretical inference, based on what we know of the sun. Hence, after the initial intellectual "gold rush", progress slows and is seriously limited. The new physics suggested by each newly discovered, active monstrosity, remains in the realm of speculation beyond the limits of direct observation. Further development requires the eventual plodding progress of hard science, where direct measurement goes hand in hand with quantitative theory. Astrophysics is, after all, an empirical science. Neither observation without quantitative theory, nor theory without quantitative observation, can progress alone. Serendipity makes a wonderful opening, but it can never play out the end game to a clear decision.

It is here that the sun has played, and continues to play, a fundamental role, providing the one area in which hard science--the critical interplay of theory and measurement--can function. It is an interesting exercise to recall the basic concepts, now routinely employed in stellar and galactic astrophysics, that originated, and were given substance, by studies of the sun, and could have originated only in the sun where direct measurement laid them bare to the inquiring mind.

Historically the list begins with the idea of gravitation and the self-confinement of a massive cloud of gas. The sun is the only star we can see, to know that it is, in fact, a self gravitating gaseous sphere. The temperature, luminosity, mass and radius of the sun had to be known before there was a firm basis for arguing that the distant stars are suns. The gaseous nature of the sun had to be established before the concept of the self-gravitating, self-supporting, luminous gaseous sphere could be developed quantitatively to give an idea on conditions in the deep interior of the sun, or other stars.

The idea of the stellar atmosphere developed only because we could see the atmosphere of the sun, particularly during eclipses. The list begins with the chromosphere and the corona, which are now an integral part of our thinking on stellar atmospheres. The realization, from Edlen's interpretation of spectroscopic studies of the sun, that the corona is an extended atmosphere at a million degrees was a revelation in its day, stirring excited and impassioned controversy. It is now routinely applied to other stars, to galaxies, and to clusters of galaxies.

The traditional spectral analysis of stars based on the assumption of local thermodynamic equilibrium (LTE) and the curve of growth was violently upset by the combined theoretical and observational studies of the solar chromosphere in the late forties and early fifties. With the high resolution available at the sun the physics has been worked out so that a proper non-LTE interpretation of stellar spectra is generally available today. Temperatures, densities, and chemical abundances of stellar atmospheres and the interstellar gas, are on a firm footing finally because of the critical theoretical and observational studies of the sun.

It will be recalled that the concept of the supersonic solar wind was established from a state of controversy only because direct measurements of the corona showed its extended high temperature, and direct detection of the wind in interplanetary space settled the question of its existence. Stellar winds are now a common concept in understanding active stellar atmospheres and mass loss. The discovery of fast and slow streams in the solar wind and the related concept of the coronal hole show the remarkable channeling that may occur in stellar winds. The effect was discovered, and can be studied, only because of direct observations of the sun and direct access to the solar wind. Indeed, the whole idea of the heliosphere, or astrosphere, is a new concept created by studies of the extension of the solar wind into interstellar space.

Sunspots, which remain an obstinate feature of the sun, have been generalized to the concept of starspots because of the variable light curves of certain M-dwarfs. The explanation in terms of star spots exists only because we see the sunspots so clearly on the sun.

The remarkable non-uniform rotation of the sun, indicating a deep internal circulation, could not have been detected, much less studied, in any other star. Yet the deep circulation has implications for most main sequence stars. The neutrino emission, intimately associated within the thermonuclear energy sources of the sun and other stars, can be studied only in connection with the sun. The puzzling negative results, so far, in the course of the efforts to detect the neutrinos suggest that there may be more to the problem of stellar interiors and thermonuclear energy than originally thought.

Solar magnetic fields make up the only sample of stellar magnetic fields whose form and behavior can be studied. Indeed, it is the observational discovery of the remarkable antics of the active solar fields that impress upon us how little we understand of the behavior of large-scale fields in the convecting gases of the sun, and, hence, of other stars, and of the gaseous disk of the galaxy, and of other galaxies. The simple explanations of twenty years ago have largely fallen by the wayside as direct precise observations and quantitative theoretical work progress toward an improved understanding of the origin of stellar magnetic fields, their buoyancy, and their activity upon arrival at the surface. The remarkable tendency for the general magnetic fields of the sun to break up into separate compressed flux tubes is a general effect that could not have been anticipated and still lacks a clear explanation. Yet it is probably common to the fields of other stars, and may contribute to their behavior. Only insofar as the behavior of fields is understood in the sun, can we venture to guess what happens in other stars.

Flares were discovered first on the sun, and virtually all that we know of flares, now recognized in M-dwarfs and possibly on a grander scale in X-ray sources, active galaxies, and quasars, is based on the theory and observations of the sun. Nucleosynthesis and spallation products are an aspect of the solar flare that can be observed and studied only because of our proximity to the sun. The corresponding effects in flares elsewhere in the universe are entirely a matter of guesswork.

The remarkable tendency of the circulation and convection in the sun to change its mode, so that solar activity comes and goes with the centuries (in addition to the familiar 11-year cycle), indicates that other stars do the same. But only the sun has been well enough observed for long enough to show the effect. Indeed the variability of the sun points out the fundamental importance of the nonuniform rotation, circulation, and convection to the behavior of all stars, and represents an outstanding and very difficult problem in solar physics at the present time.

Altogether the sun, through the last century and up to the present moment, has been the origin of most of the solid concepts that make up the foundation of modern astrophysics. With the intense interest in the gigantic activity of other stars and some of the distant galaxies, studies of the activity of the sun continue to be the principal source of hard science and new concepts in astrophysics. As a matter of fact, interest in the sun, in the solar wind, and in the magnetosphere and atmosphere of Earth has been sharpened in the last few years by variations of terrestrial weather and climate in coincidence with changes in the level of solar activity. These are both long and short term variations. In the short term it has been found that the formation of a trough (high vorticity index) over the north Pacific is inexplicably more likely to occur a couple of days after a magnetic storm. The data show, too, that troughs forming after a magnetic storm grow into more robust meteorological storms than those occurring independently of a magnetic storm. The formation of troughs also shows a correlation with sector boundaries in the solar wind, but the correlation is weak, presumably because sector boundaries are only weakly connected with magnetic storms. On a longer time scale it is a curious fact that severe drought in the high prairies in the last two hundred years has coincided in every case with the deep solar minimum at the end of each 22-year sunspot cycle. All these weather variations are baffling so far, and it is not established whether their correlation with solar activity is chance coincidence or a real physical connection. But their stubborn correlation with activity compels the physicist to examine the behavior of the magnetosphere and the atmosphere of Earth with determined care.

More recently it has come to light through the historical researches of Eddy that the familiar activity of the sun--most conspicuous as the 11 or 22-year cycle of sunspots, solar eruptions, turbulent wind, and magnetospheric storms--sometimes disappears for a century at a time. The most recent period of inactivity was 1645-1715 AD. The absence of sunspots at that time was well documented by telescopic observations. The absence of the visible, white light corona was recorded by astronomers at total eclipses of the sun.

Earlier centuries without solar activity show up clearly in
the carbon 14 record, demonstrating that the 15th century AD,
and the 4th, 7th, and 14th centuries B.C. were without activity. It appears from the statistics of the most recent 4000
years that the sun spends about one-tenth of the time in the
inactive state.

Since the magnetic field of the sun is the agent responsible for solar activity, and the magnetic field is generated
by the convection and circulation in the sun, the vanishing
of solar activity implies some change in the convection and
circulation. Indeed, recent studies of the old maps of sunspots show that the global equatorial acceleration of the sun
was more enhanced and concentrated toward the equator as the
sunspots were dying away just prior to 1645 A.D.

Now the convection and circulation is also responsible
for the delivery of heat to the surface of the sun. The
surface brightness of the sun is particularly sensitive to
vertical winds. The internal circulation and convection of
the sun is a very difficult theoretical hydrodynamic problem
that is only beginning to be studied, so we do not know the
nature of the motions for either the familiar active state,
or the 17th century inactive state, of the sun. We can only
suppose that the circulation of the sun changes from one mode
to another, as the atmosphere of Earth is observed to do from
time to time. It is difficult to conceive of a significant
change in convection and circulation that would not affect the
surface brightness of the sun in some slight degree, either
changing the total luminosity for a time or redistributing the
brightness between the poles and the equator. In either case
it is possible to check the mean annual temperature at Earth
for an effect in the solar "constant". The historical record
covers the 15th and 17th centuries very nicely, while the
mean temperatures in the earlier centuries show up in the
advance of glaciers, in the oxygen isotopic ratios, in the
abundance of various species of micro-organism skeletons in
sediment, etc. The facts are that the centuries of cold, in
which the mean annual temperature in the northern temperate
latitudes was depressed by about 1°C, occur in coincidence
with the centuries of solar inactivity, and at no other time,
in the last 4000 years. Thus it appears that the equatorial
brightness of the sun, if not the entire sun, may be depressed
by a small amount (a fraction of a percent) when the sun is
in the inactive mode.

It is an interesting fact that historians have been aware
of the cold, the crop failures, and the associated political
unrest, in the 15th and 17th centuries irrespective of any
variations of the sun. The period has been called the "little

2

Historical and Arboreal Evidence for a Changing Sun

John A. Eddy

Is the sun a constant star or does it change? And in what ways? Modern observations have not completely answered these questions but they leave little doubt that the sun is ever varying: sunspots appear and die in regular cycles, flares erupt, and prominences come and go. At every level the solar surface is in constant turmoil and the magnetic fields that rule the sun are never still. We generally assume that in the longer term, of centuries and millenia, the sun's behavior is more or less constant, but this is only a first assumption, based as much on hope as on fact or physical theory.

One of the intriguing problems of solar physics has been the testing of the premise of long-term solar constancy: the challenge to uncover, by whatever means available, the all-important history of our sun. It is as well a practical problem, for long-term solar variation could have climatic effects, and as in terrestrial affairs we can intelligently anticipate the future only when we know the past. Serious efforts began in about 1850, when, following the discovery of the sunspot cycle, Rudolf Wolf in Zürich searched historical accounts to recover sunspot numbers 150 years backward in time (1). Over this short span there seemed to be a constancy that may have long misled us. What other evidence is there, and how does one look farther into the past? For much of time we have only indirect or proxy data, as for example in the annual record kept by trees. In reading this record there have been false starts and false hopes, but with modern methods a new and consistent picture now emerges, with the promise of an even clearer one to come.

The Work of A. E. Douglass

In the early years of this century, when he was middle aged, Andrew Ellicott Douglass of Arizona took up the study of tree-rings: a self-taught hobby that became, in time, a standard tool of archaeology and climatology. The development of dendrochronology is an illuminating story -- not so much for what it tells of trees and how they grow but for what it says of science and the way that fundamental ideas unfold: from the imagination of single minds, often spreading in unanticipated directions and seldom respecting the boundaries fixed by conventional disciplines.

Douglass was an astronomer and his original interest in trees, beyond an admitted love of piney forests, was based on the hope of finding there the history of the sun. It was an heroic quest in which he never gave up hope. Historical records of direct, solar observations had established that the well-known 11-year cycle of sunspots and solar activity had been in operation since at least 1700, and possibly 1610, when the telescope was first turned on the sun. But had the sun always behaved in this way? The answer must come from proxy data. Tree-growth, Douglass demonstrated, could be used as an objective diary of local conditions through the measurement of annual growth rings. In this record, that reached far back into prehistory, Douglass hoped to find the signature of unseen solar cycles of the past.

His hope, he well knew, was based on the problematical existence of a real connection between solar activity, measured in sunspot numbers, and local weather: that somehow rainfall and other meteorological inputs to tree growth were regulated by varying activity on the sun. Were this the case, one should be able to recognize an 11-year pattern in annual tree-ring widths, reflecting the observed cycle of annually-averaged sunspot numbers.

So simple a link between tree growth and sunspots now seems naïve, but we must remember that in the early 1900's there were louder, unrefuted claims for direct solar-weather connections, and a simpler and more innocent picture of global meteorology. At about the same time that Douglass began his search in trees, Charles Greeley Abbot began a similar study at the Smithsonian Institution in Washington, launching a long program to measure the solar constant, in which he also expected to find a dominant 11-year cycle (2). At about the same time Norman Lockyer, the best known solar physicist in England, had convinced himself and others that the 11-year cycle of solar activity controlled terrestrial climate and weather trends, particularly recurrent famines and monsoons in the tropics, where, he reasoned, solar influence would be more readily felt and more easily recognized (3).

Douglass' early work taught him that tree growth was itself a tangled product of many factors, and in patiently un-
ravelling them he established the basic principles of modern
dendrochronology. In the process he became convinced that he
had found cycles in tree growth that were related to the 11-
year cycle of activity on the sun (4). In tree-ring patterns
these cycles were, admittedly, subtle things: sometimes
there, sometimes not; present, weakly in some localities, and
in others completely hidden. In the arid Southwest, where
Douglass took most of his samples, and where dendrochronology
was expected to be simpler, the anticipated solar imprint was
not a particularly dominant feature; in the giant sequoias of
California, where he secured his longest, continuous tree-
ring chronologies, there seemed to be no cyclic solar influence at all. Douglass also noted the existence of prolonged
periods, as between about 1650 and 1700, when Southwest tree-
rings were consistently narrow and when the alleged solar
signal disappeared altogether (5).

One of the clearest cases of pronounced 11-year power in
ring widths was found in a number of Scotch pines grown over
an 80-year period in a controlled forest at Eberswalde near
Berlin (Fig. 1). We can understand why this isolated example
was not uniformly convincing. Aha! detractors must have said,
you have selected an artificial case. These trees are grown
under enforced conditions, and we all know what the Prussians
are like: the trees, like the people, have learned to obey
orders! Not so, said Douglass, who had convinced himself that
there had been no enforced pattern of cultivation or watering;
these very trees, well-nourished, less subject to certain
insect scourges and the complications of crowding by older
trees, were in his view ideally selected to demonstrate potential solar influence.

Douglass used various methods to identify cyclic tree-
ring patterns, including ingenious optical devices called the
"periodograph" and the "cycloscope" that could recognize and
display periods of subtle, recurrent patterns (6). But when
he died, at 95 in 1962, the situation was still not clear.
Skeptics who looked closely at Douglass' major reports on the
subject could say that he had found only scattered cases of
accidental relationships between tree-growth patterns and the
solar cycle. Those who liked cycles, or who were otherwise
convinced of cyclic, sun-weather relationships, could, on
the other hand, select data -- like the Prussian pines -- that
bolstered their cause.

Figure 1. Section of a Scotch pine from a tended forest at
Eberswalde, Prussia, planted in about 1820 and cut
in 1912. Arrows, placed by Douglass, mark years of
maximum sunspot number, showing, in this selected
sample, an apparent correlation with maximum annual
tree growth. From Douglass (4), Vol. I, pp. 37-39,
74-76.

Recent Studies of Tree-Rings and Climate

To clarify the case, in 1972 LaMarche and Fritts, both at Douglass' Laboratory of Tree-Ring Research in Tucson, tried more objective tests, applying the power of digital computers and a variety of modern, statistical techniques to some of the same, Southwest tree-ring chronologies that Douglass had used (7). In none of their cases could they find evidence of any significant power at periods of 11-years. There were small but significant peaks at 20-29 year periods in certain of the tree-ring records, but LaMarche and Fritts could find no convincing relationship between these possible 22-year waves and the double sunspot cycle, and concluded that they were probably unrelated to the sun.

Their conclusions, based on powerful and repeatable studies of 49 tree-ring chronologies of western North America, seemed to confirm the more negative, modern consensus against a significant connection between the sunspot cycle, weather, and regional tree-growth. And there the story seemed to end.

Recently, however, Stockton and Meko of the same Tree-Ring Laboratory came almost by accident upon new evidence in tree-rings for a significant 22-year cycle that could be sun-related (8). When the western U.S. was divided into climatic regions, and years of drought identified in each as narrow tree-ring growth, Stockton and Meko found that the number of regions showing significant drought suggested a cycle of about 20 years, that has persisted since at least 1700. Their finding was not inconsistent with that of LaMarche and Fritts, for they too had seen a suggestion of power at the same frequency in certain localities. The more dominant signal in Stockton's data would follow if the apparently recurrent drought wandered over the American West in its area of maximum impact. Were this the case the solar signal, if that was what it was, could escape detection in tree-ring records from single sites.

More recently, Mitchell, Stockton and Meko subjected the Western drought record to exhaustive statistical tests, and have established that the 22-year period in recurrent Western drought is the strongest periodic signal present, closely in phase with the cycle of solar activity (9). Their tentative finding, potentially the strongest evidence yet for an important solar-weather connection, opens again the old question of a significant link between the sunspot cycle and the weather. Because of the regionally-complex nature of the Mitchell, Stockton and Meko result, however, it offers at best a slippery grasp on the opposite problem of reading solar history in tree-ring widths. If the sun modulates

drought in a way that distributes it more or less randomly
over a large area, an unknown solar signal derived from such a
pattern would have to be reconstituted from wisely selected
samples -- like piecing together a coded message from shredded
scraps scattered over a large area. History derived in this
way, though better than none, would seem far less authoritative than Douglass's early hope of finding in every tree a
redundant book.

Radiocarbon

In the meantime, a new way of reading solar history in
tree-rings had been developed: a more powerful method that
did not depend upon vagaries of local tree growth, or on
assumptions of the solar control of weather. In the 1960's,
Stuiver, Damon, Suess, and others (10) demonstrated that radiocarbon in plant cellulose could be interpreted as a gross
record of past solar activity.

Radiocarbon, or ^{14}C, is produced in the upper atmosphere
of the earth through the impact of high energy galactic cosmic
rays (11). Our receipt of cosmic rays is modulated, in turn,
by solar activity, through the action of the extended magnetic
field of the sun, that scatters or deflects a fraction of the
cosmic ray flux (12). When the sun is more active, it shields
the earth from some of the high energy cosmic rays (as established by measurements of nucleonic flux) and the production
of radiocarbon in the upper atmosphere is diminished. When
the sun is less active, as, for example, at minima in the 11-
year solar cycle, we receive more cosmic rays and radiocarbon
production is increased. Thus, if we had a record of how much
radiocarbon was in the atmosphere in the past we could in
principle deduce the state of solar activity at the time.

Trees keep that record for us, for atmospheric radiocarbon (as carbon dioxide) enters their leaves in photosynthesis
and is preserved, as cellulose, in new growth wood. Growth
rings of many temperate-latitude trees, as Douglass demonstrated, can be identified as an annual diary that extends, in
long-lived species such as the bristlecone pine, for many
thousands of years.

There are, to be sure, significant limitations. The half-
life of radiocarbon, 5730 years, limits the amount of the isotope available for analysis in very old wood, eventually
erasing the record. A more real restriction is imposed by the
atmosphere and oceans of the earth: radiocarbon is produced
high in the upper atmosphere and the trees are at the bottom.
Between, in the varying patterns of atmospheric circulation,
diffusion, and ocean absorption is a complex, low-pass filter

that delays and dilutes real variations in the radiocarbon production, severely attenuating changes shorter than about 20 years (13). As a result there are as yet no unambiguous identifications of the 11-year solar cycle in tree-ring radiocarbon. Only longer-term, gross effects get through in presently-readable form, and these are shifted in time (14). Moreover since the sun is not the only modulator of cosmic rays or radiocarbon, we expect the ^{14}C record of tree-rings to be written over with other histories, including certain anthropogenic effects, the changing strength of the earth's magnetic moment, and variations in the ocean-atmosphere filter.

A 2000-year record of deviations in tree-ring radiocarbon (^{14}C/^{12}C ratio), compiled by Damon, is shown in Figure 2 (15). We have here plotted it upside down, with increasing radiocarbon downward, to agree in sense with solar activity. Excursions, in parts per thousand, are shown relative to an arbitrary, 19th century reference level. The curve should be familiar to archaeologists, who have reason to despise it and would like to see it flat, for the ^{14}C excursions that interest us here are the very features that have necessitated troublesome readjustments in radiocarbon dates in the last few years -- illustrating the adage that one man's signal is another man's noise.

Is there evidence for solar modulation in the radiocarbon record in Figure 2? The gradual fall from left to right (increasing ^{14}C/^{12}C ratio), is, as we shall see, probably not a solar effect but the result of the known, slow, decrease in the strength of the earth's magnetic moment, exposing the earth to ever-increased cosmic ray fluxes and increased radiocarbon production (16). The sharp upward spike at the modern end of the curve, representing a marked drop in relative radiocarbon, is generally attributed to anthropogenic causes -- the mark of increased population and the Industrial Age (17). The burning of low-radiocarbon fossil fuels -- coal and oil -- and the systematic burning off of world forests for agriculture (18) can be expected to dilute the natural ^{14}C/^{12}C ratio in the troposphere to produce an effect like the one shown, although other contributions, including a real increase in solar activity, may be hidden under the rapidly rising curve. Much of the remaining structure is probably measurement error, for the curve comes from amalgamating radiocarbon results from different laboratories and widespread tree locations. Three striking features of the curve, however, are probably marks of a changing sun: an upward excursion (decreased radiocarbon) about AD 1200, indicating high solar activity, and two marked and prolonged dips, like a "W", in the direction of decreased solar activity at roughly AD 1500 and AD 1700. These distinctive dips (increased relative radiocarbon) were the first

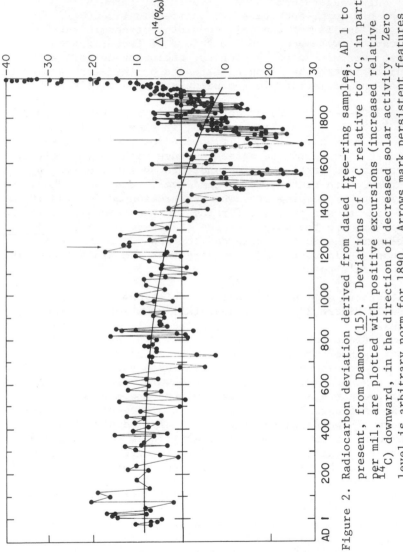

Figure 2. Radiocarbon deviation derived from dated tree-ring samples, AD 1 to present, from Damon (15). Deviations of ^{14}C relative to ^{12}C, in parts per mil, are plotted with positive excursions (increased relative ^{14}C) downward, in the direction of decreased solar activity. Zero level is arbitrary norm for 1890. Arrows mark persistent features identified as possible solar anomalies: right to left, Maunder Minimum, Sporer Minimum, Medieval Maximum.

major excursions noted in the radiocarbon history and called
the "DeVries Effect"; first found in 1958 (19) they are a consistent global feature of the tree-ring record. That the
DeVries effect could be the mark of the sun was soon pointed
out (10). But was it really solar? And if so, what did it
tell of the sun's behavior at the time? The more recent of
the dips, lasting from the middle 17th century to the early
18th, fell fortunately within the time of "modern" observation of the sun, after the introduction of the telescope,
when detailed observations of the sun began; it thus became
the key, and potential yardstick, for the interpretation of
the remainder of the radiocarbon record.

The Maunder Minimum (1645-1715)

That something unusual happened on the sun between the
middle 17th and early 18th centuries was first pointed out
by Spörer and Maunder late in the last century (20) and more
recently reaffirmed (14,21). First-hand, historical accounts
of the era tell of a surprising absence of sunspots and other
signs of solar activity, including a marked, coincident drop
in auroral reports between about 1645 and 1715.

It was a time of active interest in astronomy, and detailed interest in the sun. During the time the Greenwich
and Paris observatories were founded, there were rapid
advances in telescope technology, and the first scientific
journals appeared. Distinguished astronomers such as Flam-
steed, Halley, Cassini, and Hevelius, whose work in other
areas seems above suspicion, tell in specific terms of an
unusual absence of spots on the sun, and of looking, for as
long as 10 years, before finding any, making it seem irrespon-
sible to attribute the apparent absence to ignorance or over-
sight. Occasional spots were observed from time to time but
they were restricted, throughout the 70-year period -- through
six ordinary solar cycles -- to largely single spots and to
belts of latitude near the equator of the sun, as at the minimum of the normal sunspot cycle. We may assume, as did
Maunder, that the 11-year sunspot cycle was in operation
during the time, at a greatly reduced or submerged level, but
there is as yet no convincing evidence one way or the other
on the question, for spots and aurorae that were reported are
too infrequent for a convincing test. Eclipses of the sun
during the time were assiduously observed, but no one tells
of a structured corona, and we may presume that a coincident
absence of organized magnetic fields on the surface of the sun
greatly suppressed or even erased the corona as we see it in
the modern range of solar activity. The character of solar
surface rotation, as reconstructed from historical sunspot
drawings, seems to have altered significantly between 1625 and

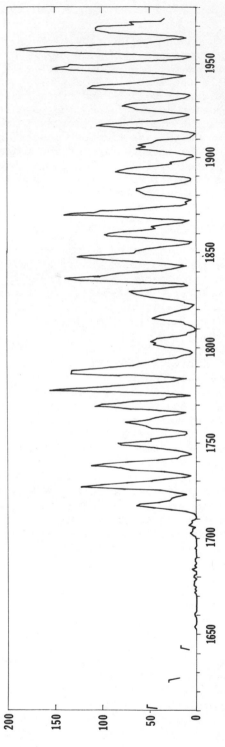

Figure 3. Annual mean sunspot number, AD 1610-1975, from Waldmeier (1) and Eddy (14), based on controlled observation from 1853 and reconstructed from less complete observations in earlier periods. Period from 1645-1715 is the Maunder Minimum.

the onset, in 1642-1644, of the Maunder Minimum (22) a finding
consistent with what is currently known of the mechanism of
operation of the normal sunspot cycle.

Thus it seems that the major feature of the radiocarbon
curve within the grasp of detailed, historical test was indeed
a time of anomalous solar activity -- an excursion at the
proper time and in the proper direction, of prolonged solar
quiet, to explain the nature of the radiocarbon anomaly. The
historical record of sunspots (Fig. 3), aurorae, solar rotation, and appearance of the sun at eclipse during the Maunder
Minimum may be adequate to provide a meaningful yardstick to
scale the solar information in the radiocarbon curve (14,23).
Further, weaker confirmation comes from pressing the same
historical records further, to the time of the two earlier
radiocarbon excursions noted in Figure 2: the earlier prolonged dip of the DeVrie Effect, about AD 1450-1540, and the
possible rise in solar activity in the 12th century. Aurora
counts, eclipse descriptions, and the far weaker record of
pre-telescope sunspot reports made with naked eye are all consistent with a solar cause for both of these two events, that
we could call the Spörer Minimum and the Medieval Maximum.
During the Medieval Maximum there were more reports of aurorae
and of naked eye sunspots than in three centuries before or
after it, and during the Spörer Minimum both indicators fell
to anomalously low levels (24).

Indirect evidence for a period of prolonged solar inactivity consistent with the Maunder and Spörer minima has been
presented recently by Forman, Schaeffer, and Schaeffer (25)
who measured concentrations of an argon isotope (^{39}A) in
recent meteorite falls. Like radiocarbon, ^{39}A on meteor surfaces is produced by high energy cosmic rays that are modulated by the extended field of the sun; in their travels through
the solar system, and before they fall beneath the protective
blanket of our atmosphere, meteors continuously sample interplanetary cosmic ray levels and thus indirectly record a time
integral of past solar activity. The half-life of ^{39}A, 269
years, makes it suitable for testing levels of solar system
cosmic ray fluxes over about the last 400-500 years. Were the
averaged level of solar activity during this time significantly lower than the present it should appear in meteor samples
as unusually high abundances of the isotope. In the 30 meteorites studied by Forman, Schaeffer, and Schaeffer, ^{39}A abundances were remarkably similar, and all were consistently high,
consistent with conditions of almost no solar modulation of
galactic cosmic rays during the period of the Maunder and
Spörer minima.

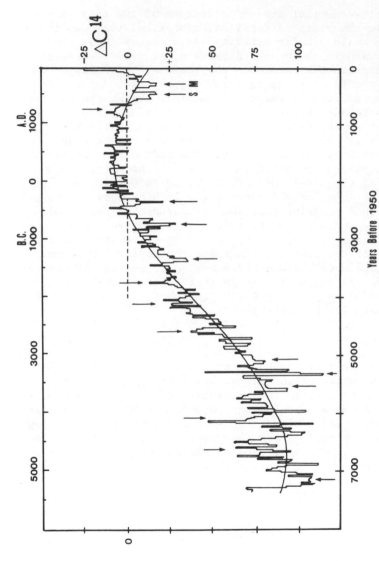

Figure 4. Radiocarbon deviation, as in Figure 2, from tree-ring samples since about 5000 B.C., from Lin, Fan, Damon, and Wallick (26). Solid curve, also from (26), represents strength of earth's magnetic moment, de-rived from paleomagnetic data by Buch (16). Features selected as possible solar excursions in Table 1 are marked with arrows.

Reading the Radiocarbon Record

With the Maunder Minimum defined and clearly identified
in the radiocarbon record, and with weaker evidence for the
reality of the two other, preceding solar effects, we can turn
to the longer record of radiocarbon history to identify other
possible features of solar cause. A compilation of tree-ring-
derived radiocarbon data by Lin, Fan, Damon, and Wallick (26)
is shown in Figure 4, spanning the remarkable range of over
7000 years, including data from trees that lived before the
dawn of the Bronze Age. At the right-hand, modern end of the
curve we recognize the same features that were identified in
Figure 2: the fossil fuel effect, the Maunder and Spörer
minima, and the Medieval Maximum. But the dominant feature
in this longer record is the possibly cyclic undulation in
overall level that reaches a minimum at about 5000 B.C. and
a maximum shortly after the time of Christ. The radiocarbon
data has been fit by Lin, Fan, Damon, and Wallick with a
smoothed curve (solid line) based on values of the earth's
magnetic moment derived from paleomagnetic data by Bucha (16).
The remarkable fit establishes the changing strength of the
earth's magnetic field as the dominant modulator of radiocarbon production.

Excursions above or below this averaged curve could be
noise, for the radiocarbon record is in a yet imperfect, early
state. They could conceivably be shorter-scale excursions in
the strength of the earth's magnetic moment (27). We may
assume, however, that some of the excursions -- and all the
major ones if we take the record of the last 2000 years as
typical -- are of solar cause. Damon has pointed out that
solar modulation can be expected to be the principal cause of
radiocarbon excursions once the smoothed, geomagnetic modulation is removed (29); he has also demonstrated (30) that the
nature of the excursions in the data of Figure 4 are consistent with what we would expect from real, external causes:
they are of greatest amplitude at the early end of the curve
when the earth's magnetic moment was at its minimum, and are
distinctly suppressed at the time of maximum field strength,
ca. AD 200, as we would expect were solar modulation being
squelched by a competing modulation of longer term exerted
by the earth's magnetic moment. Thus in interpreting the
amplitudes of any real excursions we must make correction for
this effect.

An interpretation of the long-term radiocarbon curve,
from (28), is given in Figure 5 and Table 1. In Figure 5a we
show the major excursions from the radiocarbon record of Figure 4, with a linear correction for amplitude (28) and with
the background, geomagnetic curve removed. 18 tentative

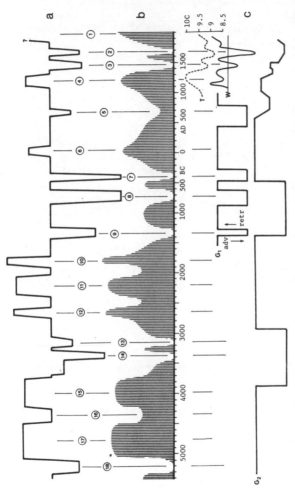

Figure 5. Interpretation of radiocarbon deviations in terms of solar effects, with climate correlation from Eddy (28). Curve (a): persistent radiocarbon deviations from Figure 4, plotted schematically and normalized to feature 2 (Maunder Minimum): downward excursions, as in Figures 2, 4 indicate increased ¹⁴C and imply decreased solar activity. Circled numbers identify features described in Table 1. Curve (b): interpretation of (a) as a long-term solar activity envelope (of possible sunspot cycle). Curve (c): four estimates of past climate. Step curve G₁: times of advance and retreat of Alpine glaciers, after Le Roy Ladurie (35); curve G₂: same, for worldwide glacier fluctuations, from Denton and Karlen (36); curve T: estimate of mean annual temperature in England (scale at right) after Lamb (34); curve W: winter-severity index (colder downward) for Paris-London area, from Lamb (33, 34).

TABLE 1

Apparent Solar Excursions in Radiocarbon Record

Feature (Fig. 5)	Beginning & End in Radiocarbon Record		Probable Extent in Real Time		^{14}C Amplitude	Corrected
1. Modern Maximum	AD 1800?	---	AD 1780?	---	?	?
2. Maunder Minimum	AD 1660	AD 1770	AD 1640	AD 1710	-1.0	-1.0
3. Spörer Minimum	AD 1420	AD 1570	AD 1400	AD 1510	-1.0	-1.1
4. Medieval Maximum	AD 1140	AD 1340	AD 1120	AD 1280	0.7	0.8
5. Minimum	AD 660	AD 770	AD 640	AD 710	-0.6	-0.7
6. Maximum	AD 1	AD 140	20 BC	AD 80	0.6	0.7
7. Minimum	420 BC	300 BC	440 BC	360 BC	-2.0	-2.4
8. Minimum	800 BC	580 BC	820 BC	640 BC	-2.1	-2.4
9. Minimum	1400 BC	1200 BC	1420 BC	1260 BC	-1.5	-1.5
10. Maximum	1850 BC	1700 BC	1870 BC	1760 BC	1.6	1.5
11. Maximum	2350 BC	2000 BC	2370 BC	2060 BC	1.4	1.2
12. Maximum	2700 BC	2550 BC	2720 BC	2610 BC	1.7	1.3
13. Minimum	3200 BC	3050 BC	3220 BC	3110 BC	-1.1	-0.7
14. Minimum	3410 BC	3270 BC	3430 BC	3330 BC	-2.8	-1.8
15. Minimum	3670 BC	3410 BC	3690 BC	3470 BC	-0.7	-0.4
16. Maximum	4220 BC	3700 BC	4240 BC	3760 BC	1.5	0.9
17. Maximum	5050 BC	4450 BC	5070 BC	4510 BC	1.7	1.0
18. Minimum	5300 BC	5050 BC	5320 BC	5110 BC	-1.5	-0.9

features have been identified, and are listed in Table 1. Their starting and ending dates have been corrected for a delay of 40 years between change in radiocarbon production and tree assimilation, derived from comparing the historical dates of the Maunder Minimum with its signature in the radiocarbon record (14,28). Amplitudes have been scaled, after making the amplitude correction described above, to unit excursion for the Maunder Minimum, which we have taken as our yardstick of solar change.

In identifying major excursions in the radiocarbon curve and labelling them as real we have accepted the fit of the solid, geomagnetic curve in Figure 4 as exact. It should be obvious that shifts in the fit of this curve, to right or left, up or down, will change the magnitude and apparent duration of some of the identified features. And we should keep in mind the imperfect nature of the radiocarbon curve. Most of the data that make up the radiocarbon curve represent averages over time intervals of about 10 years, taken by analyzing wood from many adjacent rings, to explore the general form of variations. Finer resolution, that now seems possible, should clarify and sharpen details of individual excursions, and identify the ones that are not real.

In Figure 5b we have used our Maunder Minimum yardstick to interpret the 18 radiocarbon excursions as features of a smoothed curve of changing solar activity. We may think of this shaded curve -- drawn with obvious artistic license -- as the upper envelope of a possibly-continuous 11-year cycle of solar activity, for from the time of the Maunder Minimum to the present era the radiocarbon curve has closely approximated the envelope of observed sunspot numbers (14). By this interpretation we presume that the sunspot cycle was suppressed to very low levels (as during the Maunder Minimum) at times of features numbered 2, 3, 7, 8, 9, etc, when we may assume there were few spots on the sun and few aurorae. At times numbered 4, 6, etc, the envelope of a hypothetical sunspot curve would have been unusually high. As yet, however, we can only assume the existence of an 11-year sunspot cycle in any period before about AD 1700, for there is no unequivocal evidence for its operation before that time (24).

An important question concerns the present level of solar activity, of the last 100 or 200 years, from which all our careful observations of the sun, and all our generalizations have come. Can we take it as the normal level of solar behavior -- as is usually done -- or is it anomalous, in the sense of being unusually active? We cannot find the answer directly in the radiocarbon curve, for the fossil fuel effect apparently overwhelms or masks the natural radiocarbon record

of the last century or more. There are other indications,
however, that suggest that the present era is indeed a time of
higher than usual activity on the sun (14); evidence for this
comes from a dramatic and sustained increase in aurora reports
in the early 18th century, an utter lack of descriptions of
the structured corona at eclipses before that time, and the
fact that since the end of the Maunder Minimum the envelope
of the sunspot cycle has been consistently rising. The curve
of Figure 5b has been drawn with this interpretation, and it
suggests that in the long run of thousands of years, the present levels of sunspots, flares, coronal transients, and perhaps the appearance of the corona itself -- may well be unusual, applying but a fraction of the time, and only during the
upward excursions of the long-term curve. By this interpretation features like the Maunder Minimum, of which many appear
in the radiocarbon record, may be just as common as today's
solar behavior, or perhaps moreso.

Do these first looks at solar variability from tree-ring
radiocarbon say anything of long-period, cyclic variations on
the sun? In answering the question we must be careful, for
we are again limited by the preliminary nature of the radiocarbon curve, and by possible errors in identifying major
excursions from the fitted, geomagnetic curve. Errors in this
process would probably act to hide or confuse any real, cyclic
trends. Indeed, if we take Table 1 and Figure 5 at face value
there is little evidence for any persistent, long-term cycles:
the identified long-term minima and maxima do not alternate,
but come as often in multiple sets, as in the successive Spörer and Maunder minima and other, similar features earlier in
the curve. An independent, mathematical analysis of the same
radiocarbon data by Damon (30) found no persisting long-term
cycles in the range of several hundred years. On the other
hand, if the multiple minima or maxima are grouped as single,
longer features, we see evidence of a major, long-term cycle
of about 2500 years period. A possible solar cycle of this
length has been pointed out in earlier radiocarbon data by
Bray (31), who suggested its relationship to climate changes;
it also appears as a pronounced feature in Mitchell's power
spectrum of terrestrial climate (32). In Figure 5 it shows
up most distinctly between about 5000 and 200 B.C., when two
full cycles can be identified. The combined Spörer and Maunder minima, considered as a single feature, fall at about the
right time for the next minima of a 2500-year cycle, but preceding them, in place of an expected maximum, we find a
possible, shallow depression. The missing maximum, at about
AD 400, would have occurred near the time of greatest geomagnetic squelching, when we might expect its imprint to be
diminished.

If the 2500-year cycle of long-term solar modulation is real we are today early in one of its maxima, and on this basis we might expect solar activity to continue at the level of the present era for at least several hundred years.

Conclusions

The solar variations found in tree-ring radiocarbon history are in the overall level of solar activity and surely relate to other basic changes on the sun. In the modern period of overlap with historical records of sunspots and aurorae they represent a gross modulation, as cause or effect, in the 11-year cycle of solar activity -- depressions to very low levels, as during the Maunder Minimum, and amplifications to states of prolonged high activity at other times. The modulation is slow and ponderous and could easily be missed in real-time observation; as in the case of the Maunder Minimum it can probably best be recognized in retrospect, in the peaks of recorded sunspot numbers or in compressed and integrated records such as the long history of tree-ring radiocarbon.

In earlier parts of the radiocarbon record, where we have no historical checks, the purported solar fluctuations tell us directly only of changes in the sun's modulation of cosmic rays, and therefore of changes in the solar wind and the sun's extended magnetic field. But as in the case of sunspots, these changes probably relate to other, deeper-lying changes in the sun. We have cited evidence that the nature of solar surface rotation varies significantly with long-term trends in solar activity, like the Maunder Minimum (22). This suggests, in turn, that these long-term activity variations tell of circulation changes deeper within the solar atmosphere, in the convective zone, and possibly of variations in the outward flow of radiation.

Are these long-term changes in the level of solar activity related to changes in our climate? Available evidence of climate history -- which is far from complete -- suggests strongly that they are (28). The Maunder and Spörer minima in solar behavior fall during the two most severe dips in world temperature in the last 1000 years, and their combination --the "W" in the radiocarbon record -- coincides with a more protracted period of world cold known as the Little Ice Age (33). And the modern era of climate, like the modern era of solar behavior, seems to describe a similar, possibly anomalous case: in the perspective of 1000 years, we live in a climate unusually warm and benign, and perhaps significantly, under an unusually active sun.

The correlation with the presently known history of climate change is shown in Figure 5c, from (28). European temperature and winter severity estimates of Lamb for the last 1000 years (34,33) are a remarkable fit to this best-established period of solar history. For the longer climate record, times of temperate-latitude glacier advance, derived for the Alps by Ladurie (35) and for world average by Denton and Karlén (36), match well the extended solar history curve for radiocarbon. When solar activity falls to suppressed levels like the Maunder Minimum, temperate-latitude glaciers advance; when solar activity rises to high levels, they retreat. In the global glacier record of Denton and Karlén in Figure 5 we also see the purported 2500-year period in climate oscillation, that also appears in our radiocarbon record, presumably reflecting changes on the sun.

We could be misled. It is possible that climate itself is controlling the radiocarbon content of the troposphere, in which case we should expect a 1:1 correlation in Figure 5 quite independent of the sun. This could come about through alterations in the atmosphere and ocean reservoirs that control circulation and residence times of carbon dioxide (37). Present models of the effect of temperature and pressure changes seem inadequate to distinguish between the relative amplitudes of atmospheric and solar effects. In assuming that the features we have identified are solar we can only rely on the historical record that shows the Maunder Minimum as a real solar change, unrelated to climate, on the weaker evidence for a solar Spörer Minimum and Medieval Maximum, and on the recent evidence for past low levels of solar activity from meteorites (25).

If, as is suggested, interglacial climate changes of 100 to 1000 year duration are dictated by changes on the sun, what is the mechanism that links them? It is possible that the direct solar effect measured in radiocarbon production -- the modulation of cosmic rays, and related changes in solar particle fluxes -- may be an effective trigger of climate change. The climate correlation noted by Mitchell, Stockton and Meko may add substance to such a link, for the 22-year period found in Western drought patterns points to the solar magnetic cycle as a likely cause. It could also be that activity-related changes in solar ultraviolet flux drive the climate in long term, for we know that these short-wave radiations follow solar activity (38). The ultraviolet flux received from the sun may well mimic the curve of Figure 5b, falling at times like the Maunder Minimum and rising in the present era. Were this an important, long-term climatic effect, however, we might expect a stronger correlation of climate with the 11-year sunspot cycle.

The lack of demonstrated 11-year correlations between sun and climate may be an important clue in seeking an explanation for a possible, longer-term connection. Climate seems to follow the modulation of the curve of solar activity, as read in the integrated radiocarbon record, and not the 11-year, carrier wave. This suggests that the climatically-significant changes on the sun may be the same that modulate the solar cycle, causing the depression and amplification in long-term records of sunspot numbers. The simplest mechanism for this effect could be long-term changes in the total flow of radiant energy from the sun -- the so-called solar constant -- which could depress or amplify sunspot production through the action of convective changes and the solar dynamo. By this hypothesis, were the flow of radiant energy constant, the envelope of solar activity would be more nearly so. Slow and ponderous changes in the solar constant, which would certainly be felt in climate, would in passing through the sun leave their mark by slowly modulating the long-term level of solar activity. In this case, Figure 5b could represent a history of total solar radiative flux, which by present climate models need only vary through limits of about 1% to produce climatic changes like the Little Ice Age (14). So slow and small a change, occurring over periods of about 100 years, would have escaped detection in any of our past attempts to measure changes in the solar constant and indeed would be undetectable in other stars of solar type.

As both radiocarbon and climate records improve, we can hope for a much clearer picture of their possible connection. For now, perhaps the most we can assert with assurance is that neither sun nor climate is constant. Since people are by nature poorly equipped to register any but short-term changes, it is not surprising that we fail to notice slower changes in either climate or the sun. But the trees that long outlive us remember well, keeping histories that we may only have begun to read.

Acknowledgement

I am indebted to Paul Damon for radiocarbon data, to Murray Mitchell, Charles Stockton, Minze Stuiver, and Miriam Foreman for information given in advance of publication, and to Bryant Bannister, Peter Gilman and Robert Noyes for help and discussion. My research on this problem has been supported by the Geophysical Monitoring for Climate Change Program of NOAA, and the Langley-Abbot Program of the Smithsonian Astrophysical Observatory, where I was a visiting fellow in 1977.

References and Notes

1. M. Waldmeier, The Sunspot-Activity in the Years 1610–1960 (Schulthess, Zurich, 1961), p. 8.

2. C. G. Abbot, Sun's Variation and Weather (Smithsonian Institution, Washington, D.C., 1967).

3. A. J. Meadows, Science and Controversy, A Biography of Sir Norman Lockyer (The MIT Press, Cambridge, Mass., 1972), p. 125.

4. A. E. Douglass, Climatic Cycles and Tree Growth (Publication 289, Carnegie Institution of Washington, Washington, D.C.) Vol. I, (1919); Vol. II (1928); Vol. III (1936).

5. ibid, Vol. I, p. 102; Vol. II., p. 125–126.

6. ibid, Vol. I, p. 92–97; Vol. II., p. 42–50; Vol. III., p. 40–50.

7. V. C. LaMarche, Jr. and H. C. Fritts, Tree-Ring Bulletin, 32, 19 (1972).

8. C. W. Stockton and D. M. Meko, Weatherwise, 29, 245, 1975.

9. J. M. Mitchell, Jr., C. W. Stockton and D. M. Meko, Science, in press (1977).

10. M. Stuiver, J. Geophys. Res. 66, 273 (1961); Science 149, 533 (1965); J. R. Bray, Science 156, 640 (1967); P. E. Damon, Meteorol. Monogr. 8, 151 (1968); H. E. Suess, J. Geophys. Res. 70, 5937 (1965); Meteorol. Monogr. 8, 146 (1968).

11. R. E. Lingenfelter, Rev. of Geophysics 1, 35 (1963).

12. J. A. Simpson and J. R. Wang, Astrophys. J. 161, 265 (1970).

13. P. E. Damon, A. Long, and E. I. Wallick, Earth and Plan. Sci. Letters 20, 300 (1973).

14. J. A. Eddy, Science 192, 1189 (1976).

15. P. E. Damon, private communication (1975).

16. V. Bucha, Nature 224, 681 (1970); in Radiocarbon Varia-
    tions and Absolute Chronology, Nobel Symposium 12,
    I. U. Olson, Ed. (Almquist and Wiksell, Stockholm,
    1970) p. 571–593.

17. H. E. Suess, Proc. Williams Bay Conf., Sept., NAS–NSF
    Publ., p. 52 (1954).

18. M. Stuiver, Science in press (1977).

19. H. DeVries, Proc. K. Ned. Akad. Wet. B 61 (no. 2), 94
    (1958).

20. F. W. G. Spörer, Vierteljahrsschr. Astron. Ges. (Leipzig)
    22, 323 (1887); Bull. Astron. 6, 60 (1889); E. W.
    Maunder, Mon. Not. R. Astron. Soc. 50, 251 (1890);
    Knowledge 17, 173 1894; J. Br. Astron. Assoc. 32,
    140 (1922).

21. J. A. Eddy, Sci. Am. 236, 80 (1977).

22. J. A. Eddy, P. A. Gilman, and D. E. Trotter, Solar Phys.
    46, 3 (1976); Science, in press (1977).

23. J. A. Eddy, in Physics of Solar Planetary Environments
    (Proc. Int. Symp. on Solar Terrestrial Physics,
    Boulder, D. J. Williams, Ed., Publ. by Amer. Geophys.
    Union Vol. 2, p. 958 (1976).

24. J. A. Eddy, in The Solar Output and its Variation, O. R.
    White, Ed. (Univ. of Colo. Press, Boulder, 1977).

25. M. A. Forman, O. A. Schaeffer, and G. A. Schaeffer,
    Geophys. Res. Letters, in press (1977).

26. Y. C. Lin, C. Y. Fan, P. E. Damon, and E. J. Wallick,
    14th Int. Cosmic Ray Conf., Munchen, 3, 995 (1975).

27. The nature of the auroral anomalies accompanying the
    Maunder Minimum, the Spörer Minimum, and the Medieval
    Maximum make it unlikely that any of these features of
    the radiocarbon curve was the result of short-term
    excursions in the earth's magnetic moment (28).

28. J. A. Eddy, Climatic Change, in press (1977).

29. P. E. Damon, in Scientific Methods in Medieval Archae-
    ology, R. Berger, Ed. (Univ. of Calif. Press,
    Berkeley, 1970) p. 167.

30. P. E. Damon, in The Solar Output and Its Variation, O. R. White, Ed. (Univ. of Colo. Press, Boulder, 1977).

31. J. R. Bray, Nature 220 672 (1968); Science 171, 1242 (1971).

32. J. M. Mitchell, Jr., Quaternary Res. 6, 481 (1976).

33. W. L. Gates and Y. Mintz, Understanding Climate Change, National Acad. of Science, Washington D.C. (1975).

34. H. H. Lamb, Climate: Present, Past, and Future, Vol. 1, (Metheun, London, 1972).

35. E. Le Roy Ladurie, Histoire du Climat depuis l'an mil (Flammarion, Paris, 1967).

36. G. H. Denton and W. Karlén, Quaternary Res. 3, 155 (1973).

37. P. E. Damon, in Radiocarbon Variations and Absolute Chronology, Nobel Symposium 12, I. U. Olsson, Ed. (Almquist and Wiksell, Stockholm, 1970), p. 571.

38. D. F. Heath and M. P. Thekaekara in The Solar Output and Its Variation, O. R. White, Ed. (Univ. of Colo. Press, Boulder, 1977).

3

Neutrinos from the Sun

Raymond Davis, Jr. and John C. Evans, Jr.

The sun radiates an enormous amount of energy, and we
know it has radiated energy at essentially the same rate for
billions of years. Only relatively recently has the mechanism for this energy production been understood. In 1929
d'Atkinson and Houtermans pointed out that nuclear reactions
could take place at the extremely high temperatures existing
in stellar interiors. About 10 years after this basic idea
was suggested, Bethe and von Weizsäcker in their now famous
papers (1) presented detailed mechanisms based upon the state
of knowledge of nuclear physics at that time. In the 1940's
a rapid development in the field of nuclear reactions in
stars began. An important landmark in this development was
the 1957 review paper of Burbidge, Burbidge, Fowler, and
Hoyle (1) which presented a set of processes for the synthesis of the elements by thermal fusion reaction and by explo-
sive nuclear synthesis in collapsing stars. We would like to
describe a single experiment in this grand scheme which was
designed to observe directly the nuclear processes that are
now occurring in our sun.

Solar Energy and Neutrinos

The sun is believed to produce energy by a series of
thermal fusion reactions in which four hydrogen nuclei are
converted to helium by the overall reaction

$$4H \longrightarrow {}^{4}He + 2e^{+} + 2\nu + \text{gamma radiation.} \qquad (1)$$

The detailed step-by-step series of reactions that are
presently considered to be occurring in the sun is listed in
Table I. The dominant chain of reactions generally believed
to occur in the present sun is the proton-proton chain (P-P).
The carbon-nitrogen chain (CN) plays only a minor role in the
present sun, producing only about 2 percent of the sun's
energy. The CN reactions will become more important when the

Table I

Thermal Fusion Reactions Occurring in the Sun

The P-P Chain

$$^1H + {}^1H \rightarrow {}^2D + e^+ + \nu \text{ (99.75\%)} \quad \boxed{E_\nu = 0\text{--}0.42 \text{ MeV (H--H)}}$$

$$^1H + {}^1H + e^- \rightarrow {}^2D + \nu \text{ (0.25\%)} \quad \boxed{E_\nu = 1.44 \text{ MeV} \quad \text{(PeP)}}$$

$$^2D + {}^1H \rightarrow {}^3He + \gamma$$

(95%)   (5%)

$$^3He + {}^3He \rightarrow {}^4He + 2\,{}^1H \qquad {}^3He + {}^4He \rightarrow {}^7Be + \gamma$$

$$^7Be + e^- \rightarrow {}^7Li + \nu \quad \boxed{E_\nu = 0.86 \text{ MeV}} \quad {}^7Be + {}^1H \rightarrow {}^8B + \gamma$$

$$^7Li + {}^1H \rightarrow 2\,{}^4He \qquad {}^8B \rightarrow {}^{8*}Be + e^+ + \nu \quad \boxed{E_\nu = 0\text{--}14 \text{ MeV}}$$

$$\longrightarrow 2\,{}^4He$$

The CN Cycle

$$^{12}C + {}^1H \rightarrow {}^{13}N + \nu$$

$$^{13}N \rightarrow {}^{13}C + e^+ + \nu \qquad \boxed{E_\nu = 0\text{--}1.20 \text{ MeV}}$$

$$^{13}C + {}^1H \rightarrow {}^{14}N + \gamma$$

$$^{14}N + {}^1H \rightarrow {}^{15}O + \gamma$$

$$^{15}O \rightarrow {}^{15}N + e^+ + \nu \qquad \boxed{E_\nu = 0\text{--}1.74 \text{ MeV}}$$

sun becomes much older and its central temperature increases.
The particular reactions selected are based upon extensive
laboratory studies of nuclear reactions. All but one of
these reactions have been studied at low energies, and the
reaction cross-sections extrapolated to the extremely low
energies (5-20 keV) that correspond to the interior temperatures of stars. Note especially that the basic H-H reaction
and its electron capture branch, the PeP reaction, cannot be
measured in the laboratory and must be calculated theoretically.

It is important to test our ideas about these nuclear
processes in stars. From our theoretical understanding of
these processes and of stellar structures there has been
developed a consistent set of models that explain the evolution of stars and their distribution on the luminosity-
surface temperature diagram (Hertzprung-Russell diagram)(1).
Our knowledge of stars is derived from observing the light
from their photospheres. We have no direct observations of
the nuclear processes occurring in their interiors. The only
hope of observing nuclear processes directly is to observe
the neutrino (ν) radiation from the sun. This possibility
depends upon the unique penetrating power of this massless
or nearly massless neutral particle. The properties of neutrinos are sufficiently well understood for us to be certain
that their interactions with electrons and nuclei are small
enough that they would easily escape from the interior of the
sun without loss of energy. We may note from Table 1 that in
the P-P chain there are four neutrino producing reactions;
the H-H reaction, the PeP reaction, and the radioactive
decays of ^{7}Be and ^{8}B. The neutrinos from all of these reactions have low energies except those from the decay of ^{8}B.

The characteristic weak interaction of neutrinos that
allows them to escape from the interior of the sun also makes
them difficult to observe. Neutrino interactions have only
been observed near a nuclear reactor with a high power level
and in concentrated beams of energetic neutrinos produced by
high energy accelerators. Observing the relatively low
fluxes of low energy neutrinos from the sun requires a very
large detector which must be well shielded from cosmic rays
and other nuclear particles. We will describe a radiochemical detector that is based upon the neutrino capture reaction,

$$\nu + \,^{37}Cl \longrightarrow \,^{37}Ar + e^-. \tag{2}$$

This reaction is the inverse of the normal radioactive decay
of ^{37}Ar that occurs with a half-life of 35 days. The detector was built during the period 1964-1967 with the objective
of observing the calculated flux of neutrinos, or of search-
ing for a flux 1/10th of the calculated value. It is well

known that the expected flux was not observed and this finding
has led to a careful reexamination of the theory. Because of
the theoretical interest in the results the experimental
observations have continued up to the present. During these
9 years of operation improvements in neutrino detection sensitivity have been made, and many experimental tests have been
performed.

We will describe how this experiment works, and give the
latest results. We will also review some of the ideas being
currently discussed as possible explanations for the low neutrino flux from the sun. Finally, we will present arguments
for the urgent need for other detectors of solar neutrinos,
review some possible detectors suggested in the literature,
and describe briefly our work at BNL on two radiochemical
detectors.

The Brookhaven
Solar Neutrino Experiment

Neutrino Detection

As mentioned earlier, a neutrino can be absorbed in a
nucleus by a process that is the inverse of a nuclear process that creates a neutrino. A simple example is the one
already mentioned in equation (2), in which the electron-
capture radioactive decay of ^{37}Ar is inverted. The neutrino
capture cross-section, which depends strongly upon the
nuclear structure of the two nuclei involved, can vary over
many orders of magnitude for different isobaric pairs of
nuclei. For observing solar neutrinos one must select a
favorable case, with a low energy threshold and a favorable
capture cross-section. Then one is faced with the difficulty
that the rate of solar neutrino capture is extremely low, and
background processes induced by cosmic rays and environmental
radioactive contaminants become an extremely serious problem.
A solution to the problem of low rate, and background effects,
is to use a radiochemical technique. In this approach one
can use a large number of target atoms ($\sim10^{30}$) and chemically
separate a few atoms of a radioactive product. By doing this
one sacrifices observing individual neutrino captures and any
information about the neutrino energy and direction.

The neutrino capture reaction by ^{37}Cl, which is particularly favorable, was suggested many years ago independently
by Bruno Pontecorvo and by Luis Alvarez (2). One of its main
advantages is that the product, ^{37}Ar, is a rare gas that may
be removed easily from a chlorine-containing liquid by
purging with helium gas. A convenient chlorine-containing
liquid is perchloroethylene, C_2Cl_4, an inexpensive solvent

that is used industrially for dry cleaning clothes. The other important advantage is that ^{37}Ar decay events can be easily characterized. The ^{37}Ar counting techniques that we use will be described later on. A radio-chemical experiment, of course, measures the total capture rate of neutrinos from all of the neutrino producing reactions in the sun that have an energy above the threshold, in this case 0.814 MeV for ^{37}Cl. The neutrino capture cross-section for ^{37}Cl has not been measured experimentally, but can be calculated accurately as a function of the neutrino energy. Because of the great interest in the solar neutrino problem, John Bahcall in 1964 calculated the neutrino-capture cross section for ^{37}Cl for all of the neutrino producing reactions in the sun.

Since that time he has refined his calculations and several other groups have made similar calculations (see reference 3). These cross-sections are listed in Table II along with the neutrino fluxes calculated for the standard model of the sun. The pertinent quantity for solar neutrino detection, the product of flux and cross-section, is listed for each reaction in the last column. Note that the abundant flux of low-energy neutrinos from the H-H reaction is below the energy threshold. The importance of this fact will come into our discussion later. Also, note that the low flux of the energetic neutrinos from ^{8}B decay produces the largest signal. This is an interesting result of two effects. First, the neutrino capture cross-section increases approximately with the square of the excess neutrino energy above threshold. Secondly, the ^{8}B decay neutrinos have sufficient energy to populate the analog state, an excited state at 4.9 MeV in ^{37}Ar. This transition is super-allowed and it has a much greater probability than those to all other states in ^{37}Ar combined.

The sum of the product of flux times cross-section ($\Sigma\phi\sigma$) for all neutrino sources in the sun gives us the capture rate of 5.8 x 10^{-36} captures per second per ^{37}Cl atom. We will define a solar neutrino unit, abbreviated SNU, as 10^{-36} captures per second per ^{37}Cl atom and use this convenient unit in the future discussion. In 1964 when we decided to build the Brookhaven Solar Neutrino Experiment a 100,000 gallon (3.8 x 10^{5} liters) tank was chosen. A tank of this size gave a reasonable neutrino capture rate of about 5 per day based on the 1965 theory. Today's standard model forecast is 1.2 neutrino captures per day, a rate corresponding to 5.8 SNU. Based upon this production rate and the known decay rate of ^{37}Ar there should be 52 atoms of ^{37}Ar present in the liquid in a 100,000 gallon tank after an exposure of 100 days.

Table II

Solar Neutrino Fluxes[*] and Cross Sections[**]

$$\nu + {}^{37}Cl \longrightarrow {}^{37}Ar + e^-$$

Neutrino Sources and Energies in MeV	Flux on Earth ϕ in cm^{-2} sec^{-1} (from standard solar model)	Cross Section σ in cm^2	Capture Rate in ${}^{37}Cl$ $\phi\sigma \times 10^{36}$ sec^{-1} SNU
H+H → D+e$^+$+ν (0–0.42)	6.1×10^{10}	0	0
H+H+e$^-$ → D+ν (1.44)	1.5×10^8	1.72×10^{-45}	0.26
^{7}Be decay (0.86)	3.4×10^9	2.9×10^{-46}	0.99
^{8}Be decay (0–14)	3.2×10^6	1.35×10^{-42}	4.32
^{15}O decay (0–1.74)	1.8×10^8	7.8×10^{-46}	0.14
^{13}N decay (0–1.19)	2.6×10^8	2.1×10^{-46}	0.05

$\Sigma\phi\sigma$ (total) = 5.8 SNU[***]

[*]   J. N. Bahcall, W. F. Huebner, N. H. Magee, Jr., A. L. Merts, and R. K. Ulrich (March 1973).(7)

[**] J. N. Bahcall (1966).(8)

[***]Bahcall's 1977 (unpublished) cross sections reduce this to 4.7 SNU. Model of T. R. Carlson, D. Ezer, and R. Strothers (7) with new opacities give 9.8 SNU.

Extraction of Argon

These few atoms can be efficiently removed by purging the tank with helium gas. To accomplish this we use a system of two 500 gal/min pumps and a set of 40 eductors (gas liquid nozzles). After leaving the tank, helium gas is circulated through a condenser at $-40^{\circ}C$, and then through a molecular sieve adsorber for removing perchloroethylene vapors. Then the helium gas passes through a bed of charcoal cooled with liquid nitrogen $(-196^{\circ}C)$ to remove argon. The helium is then returned to the tank. This system essentially completely removes the argon dissolved in the liquid. The recovered argon is purified by gas chromatography and placed in a proportional counter to observe the ^{37}Ar decay. To measure quantitatively the argon recovery yield, a small measured volume of ^{36}Ar or ^{38}Ar ($\sim$0.1 cm^3 at STP) is dissolved in the liquid prior to exposure. The final sample of argon is analyzed volumetrically and isotopically to determine the yield for each experiment. Experiments have shown that the carrier argon gas does indeed dissolve in the liquid. Another test was made of the recovery efficiency by producing ^{37}Ar directly in the liquid by a neutron source. This was accomplished using a re-entrant tube reaching to the center of the tank into which a radium-beryllium neutron source could be placed. It was shown that the ^{37}Ar made by this source by (n,p) followed by (p,n) reactions was removed with the same efficiency as the carrier argon. In another test 500 atoms of ^{37}Ar were placed in the tank, and then recovered. These tests showed that if ^{37}Ar were produced in the 100,000 gallon tank of perchloroethylene by neutrino capture it would be efficiently recovered.

Since the rate of production of ^{37}Ar in the tank by solar neutrinos is so low, it is necessary to shield the tank from cosmic rays by building the tank as deep underground as possible. Fortunately, the Homestake Mining Company allowed us to use their deep gold mine at Lead, South Dakota. They designed and excavated a special cavity in their mine 4850 feet below the surface to house the experiment. The general arrangement underground is shown in Fig. 1. An important feature of the arrangement is that the cavity containing the tank can be flooded with water which shields the tank from fast neutrons that are produced in the rock wall by spontaneous fission of uranium, and (α,n) reactions. The fast neutron background effect from Homestake rock is small, and is essentially completely removed by the water.

Counting ^{37}Ar

The purified argon removed from the tank is placed in a small low-level proportional counter. By using a small

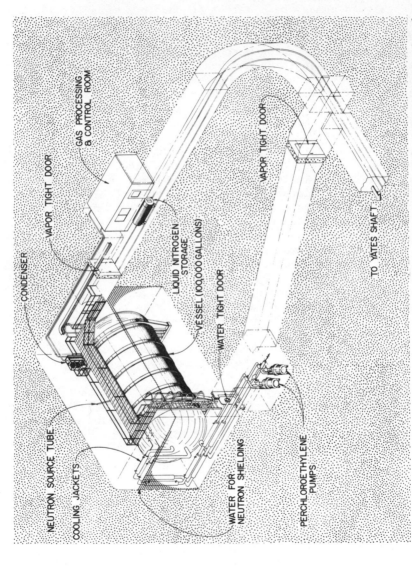

NEUTRON SOURCE TUBE

COOLING JACKETS

CONDENSER

VAPOR TIGHT DOOR

GAS PROCESSING
& CONTROL ROOM

LIQUID NITROGEN
STORAGE

VESSEL (100,000 GALLONS)

WATER TIGHT DOOR

VAPOR TIGHT DOOR

WATER FOR
NEUTRON SHIELDING

PERCHLOROETHYLENE
PUMPS

TO YATES SHAFT

Figure 1. The arrangement of the Brookhaven Solar Neutrino Experiment in the Homestake Gold mine. The apparatus shown is in a specially designed cavity in the mine 4850 feet below the surface.

counter (2.5 cm long and 0.4 cm diameter), a low background counting rate is achieved. The counter is shielded by a 20 cm thick mercury shield and operated in anti-coincidence with a surrounding large sodium iodide scintillation counter to reject counts produced by cosmic rays and environmental gamma rays. The ^{37}Ar decay events produce a group of low energy Auger electrons with a total energy of 2.8 keV. The range of these Auger electrons is extremely small in the counter gas resulting in a tiny cluster of ions that ultimately give a pulse with a fast rise-time. Thus, by measuring the pulse size and the pulse rise-time one can characterize ^{37}Ar decay events and distinguish them from events produced by various background processes. There is, of course, a small counter background of ^{37}Ar-like events, but this amounts to one count in 35 days in our best counters. One can further character-ize ^{37}Ar decays by noting whether the counts observed above the counter background decay with a 35 day half-life. The samples from individual runs were usually counted for periods of 150 days or longer. We will not present the detailed counting data here (see reference 4), but will note that in most runs there are only a few counts above counter background in a 70-day counting period. In a few experiments between 5 and 10 counts were observed above counter background for the 70-day counting period.

Results

The data for a series of 18 experimental runs covering the period April 1970 through January 1976 are shown in Fig. 2. The ^{37}Ar production rates shown have rather large errors because of the small number of observed counts. Notice that a few experiments have higher rates. An outstanding one is run 27 that was exposed over the period July 7-Nov. 5, 1972. There is no satisfactory explanation for the variations other than statistical variations. The average ^{37}Ar production rate in the 100,000 gallon tank derived from these 18 experi-ments is 0.32 ± 0.08 ^{37}Ar atoms per day. There is a back-ground production of ^{37}Ar from the cosmic ray muon flux even at the great depth of the mine. It amounts to 0.08 ± 0.03 ^{37}Ar atoms per day. Upon subtractiong this background effect one obtains a net signal that could be ascribed to solar neu-trinos of 0.24 ± 0.09 ^{37}Ar atoms per day, or 1.3 ± 0.4 SNU. We do not regard this result as a measurement of the solar neutrino flux because of uncertainties in various background effects, but do conclude that an upper limit (1σ) to the solar neutrino flux is 1.7 SNU. It is conceivable that there are variations but most theorists agree that it is extremely unlikely that the solar neutrino flux would vary, and for the present we regard these variations to be statistical. Of the 18 experiments 12 are very low, essentially at the cosmic ray

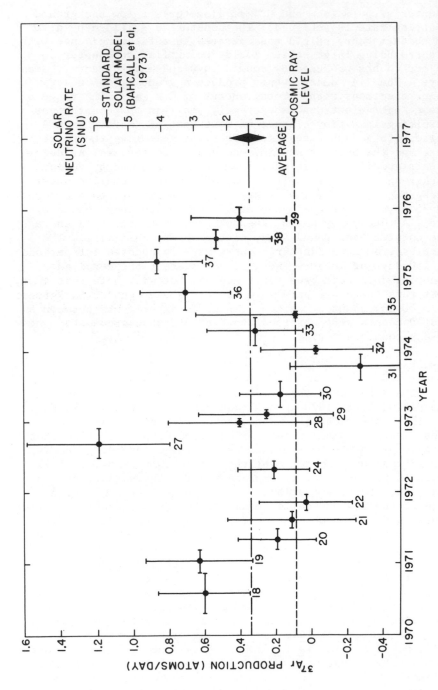

Figure 2. Summary of results during the period 1970-1976 from the Brookhaven Solar Neutrino Experiment.

background level. We would like to point out that the water
shield was not in place for experiment nos. 18, 19, 20, 38,
and 39. However, the fast neutron background is small and our
estimate of this background effect is about 0.04 ^{37}Ar atoms
per day. However, this correction was not applied.

Following run no. 39 we decided to perform a series of
ten runs in which the tank was exposed for only 35 days to
search for possible fluctuations. This series of runs is
still in progress and will be reported at a later time. In
the experiments reported above the counting measurements themselves were made at Brookhaven. However, in the near future
the counting facility will be moved underground and we hope
that there will be a reduction in our counter background.

Comparison with Solar Models

The Standard Solar Model

The results of the solar neutrino experiment are usually
compared with the so-called standard model of the sun (1,5,7).
We will now review some of the basic assumptions and physical
principles used in developing this model. One premise upon
which it is based is that the material of the sun was thor-
oughly mixed at an early stage in its development, when the
young sun was highly convective. Thus, at an early stage the
sun had a uniform composition that is taken to be identical
with its present surface composition, namely 78 percent hydrogen, 20 percent helium and 2 percent heavier elements--mostly
carbon, nitrogen and oxygen. The composition of the initial
sun determined the opacity of the material for the escape of
radiation and this in turn regulated the temperature gradient
in its interior. The composition in the interior changed as
the sun evolved and hydrogen was continuously converted into
helium. The sun is considered to be spherical and rotating
at a rate small enough that centrifugal forces are unimportant.
It is further assumed that magnetic fields are too small to be
important.

The structure of the sun is derived from a set of differential equations. The equation for hydrostatic equilibrium
expresses the fact that the gravitational force toward the
center is balanced by the outward pressure. A second equation relates the luminosity to the nuclear energy production
rate, and a third equation relates the escape of radiation to
the important opacity parameter. These equations are solved
by iterative procedures to match the present mass, age, and
luminosity of the sun after a period of evolution of 4.7
billion years.

A large amount of experimental data and theoretically derived information is introduced into these calculations. Nuclear reaction cross-sections are derived from laboratory measurements and fitted to the presumed correct theoretical form for Coulomb and nuclear barrier penetration to express the cross section as a function of the energy (6). As mentioned earlier, the cross section of initial H-H reaction must be calculated theoretically, but it is claimed that the errors are not more than 5 percent. The kinetic energy of interacting nuclei is related to the Maxwell-Boltzmann distribution of velocities, and an ideal gas equation of state is used. The opacities are calculated for the elemental composition corresponding to that observed in the present photosphere of the sun adjusted for changes resulting from nuclear burning. There are no experimental tests for the opacity calculations at the temperature and pressure conditions in the interior of stars. If processes are neglected or not recognized in the calculation, then the calculated opacity values would be too low.

The neutrino fluxes calculated from the standard solar model are given in Table II. There is some latitude in many of the parameters used and the error in the total flux-cross section product is generally regarded to be about 30 per-cent. The value predicted by the standard model for ^{37}Cl is 5.8 SNU, a value to be compared to the 1σ upper limit of 1.7 SNU given earlier. The discrepancy between the theoretically predicted value and the experimental measurements has existed for many years and there now exists a rather extensive literature discussing the problem and offering a variety of explanations. We would like to mention some of the suggestions that have been made and refer the reader to review articles for a more complete summary (8).

Models with Low Neutrino Fluxes

First, let us discuss some variations in the solar model itself that could lead to low neutrino fluxes. The emphasis has been to develop models with low fluxes to meet the demands of the experimental results. We notice that processes or effects that would lead to higher neutrino fluxes are in general not given serious consideration. Since the ^{37}Cl experiment is particularly sensitive to the ^{8}B neutrino flux and the reaction producing ^{8}B is strongly temperature dependent, various models have been tried that give reduced interior temperatures. A model was examined by Bahcall and Ulrich (9) in which it was assumed that the interior of the sun was essentially devoid of heavy elements. Because of the reduced opacity of a pure hydrogen-helium mixture this model gave lower interior temperatures. The ^{8}B flux was reduced by a

factor of 6 and the total neutrino capture rate was reduced
to 1.4 SNU. This is an attractive solution to the problem of
low neutrino fluxes, but one which requires an explanation of
the presence of the heavy elements in the photosphere of the
present sun. One possible explanation is that they appeared
after the sun had passed through its highly convective phase,
and are now present only in the outer convective zone of the
sun. They could have accumulated through infall of solar
system material, or they could have been collected by the sun
during its passage through clouds of material as it traverses
the galaxy.(9).

Lower interior temperatures have been obtained for models
that allow continuous or intermittent mixing of the interior
of the sun. It is interesting to note that if a model is
developed with the extreme and arbitrary assumption that the
sun is continuously mixed throughout its life, the neutrino
capture rate is only reduced to 2 SNU. If the sun is mixed
periodically, the neutrino flux would be temporarily depressed and the solar luminosity would exhibit long period
variations. Such luminosity variations would give a natural
explanation for glaciation on the earth (10). The difficulty
with these models is that no satisfactory mechanism has been
proposed which allows the sun to become temporarily unstable
and mix in a natural way. In fact, the interior of the sun
is regarded to be extremely stable, even though the solar
surface exhibits an 11 year activity cycle, emits intense
solar flares on occasion, shows differential rotation as a
function of latitude, and displays other surface phenomena
discussed by other speakers at this symposium.

It is possible to reduce the solar interior temperatures
by invoking either rapid rotation of the central regions or
intense magnetic fields. Solar models have been calculated
using both of these effects (11). If the sun rotates rapidly,
a centrifugal term enters the equation of hydrostatic equilibrium. These models do have lower central temperatures and
correspondingly lower ^{8}B neutrino fluxes. However, reducing
the ^{8}B flux below 1 SNU requires a very rapidly rotating
central core in the sun at the present time. These models
with rapid rotation would require the sun to be an oblate
spheroid with an oblateness exceeding that known from direct
measurement. We know from the measurements of Henry Hill and
his associates that the sun is a nearly perfect sphere with
a polar to equatorial diameter difference of 1 part in 10^5.
Since central magnetic fields exert a pressure, models
assuming intense fields of the order of 10^8 gauss also give
lower central temperatures. However, it is unlikely that
magnetic fields of this magnitude would be stable for long
periods of time without some means of sustaining them.

Recently Snell, Wheeler, and Wilson suggested that it is
possible to have the sun remain spherical in the presence of
both a magnetic field and a rapidly rotating core. Their
models can give neutrino capture rates as low as 0.6 SNU. An
important solar problem that has been discussed for many years
is that the sun has a very low apparent angular momentum based
upon the observed rotation of its photosphere. Its apparent
angular momentum is much lower than the total orbital angular
momentum of the planets. The sun does lose angular momentum
continuously by the emission of solar wind material from its
surface. There has been a great amount of discussion in the
literature about the possibility that the sun's angular momentum is higher than that deduced from its surface period. From
this viewpoint solar models with rapidly rotating cores have
considerable general interest.

Solar Luminosity

All solar model calculations follow the evolution of the
sun with time, and are normalized to the present age, mass,
and luminosity. They show that the sun's luminosity increases
with time, so that in the past the sun was less luminous than
it is now. In fact Rood and Newman (12) have pointed out that
essentially all solar models have the common result of predicting a 5 percent linear increase in luminosity every 10^9
years. This would indicate that the sun was 10 percent less
luminous 2 billion years ago. Such a lower solar luminosity
would produce a dramatic lowering of the terrestrial temperature. In fact, it is claimed that if the sun's luminosity
were to drop as little as 3 percent the oceans would freeze
solid in about 10^7 years. Since we know this has not occurred
and since it is difficult to imagine that the luminosity has
not increased with time as the solar models demand, there is
a serious lack in our understanding of the factors contributing to the terrestrial climate. The history of the earth's
surface temperature has been determined by studying the oxygen
isotope ratios in fossils. The oxygen isotope ratio is primarily controlled by the temperature of the environment when
the fossils were formed. Much of our knowledge of terrestrial temperatures in the past comes from these studies. It is
interesting that Knauth and Epstein found from studies of
oxygen isotopes in cherts formed 1.3 and 3 billion years ago
that the earth's surface attained temperatures as hot as 50
to 70°C (12). If this interpretation is correct, then one must
explain a high surface temperature of a few billion years ago,
not a low one! It is possible that the earth's atmosphere had
a different composition that in some way increased the green-
house effect. Another possibility that has been recently suggested by Turcotte, Cisne, and Nordmann is that the moon
could produce tidal heating of the earth when it was closer to

the earth a few billion years ago (12). One might conclude
that there are a large number of factors that could contribute
to the earth's surface temperature besides the solar luminosity. It could well be that the results of the solar model
calculations are not only correct, but also the best known of
the various factors.

The lack of quantitative agreement between the experimental observations and the standard model calculations has been
a problem for almost 10 years. The neutrino fluxes calculated
presently are much lower than those calculated ten years ago.
The major changes arise from new laboratory and observational
data introduced into computations. Although it contains many
oversimplifications, the standard model still stands as the
one most likely to be correct.

Some Special Models

There are a number of radical approaches to the problem
that should be mentioned (13). Hoyle suggested that the interior core of the sun is highly enriched in elements heavier
than magnesium that come from a condensation of debris from
an early cosmic event. Hydrogen and helium condensed later
on the surface of this dense core to form the present sun.
Others have postulated that the sun has a very helium-rich
core. These models with heavy element or helium cores allow
the hydrogen fusion reactions to take place in regions outside
the core where the temperature is lower, and thereby produce
low ^{8}B neutrino fluxes. It has also been suggested by
Clayton, Newman, and Talbot that the sun derives part of its
energy from a black hole in its center. Special models of
the sun have also beeen constructed that postulate a change
in the gravitational constant with time. Although these and
many other similar suggestions could be correct and their
consequences should be explored, we feel that the explanation
for the low solar neutrino fluxes will turn out to be within
the framework of the present ideas of stellar structure and
hydrogen fusion.

Neutrino Properties

Finally, we would like to mention that the Brookhaven
solar neutrino experiment depends also upon the present theory
of weak interactions being correct, and in particular, upon
the properties of neutrinos derived from this theory. The
important properties are: (1) the neutrino interacts with
matter, electrons and nuclei, only by known scattering and
absorption processes that are reliably calculated by the
theory, and (2) the neutrino does not decay or change into
another particle with a different absorption cross-section.

Theories have been proposed with neutrino properties different from those assumed above (14). One possible neutrino
property that has been discussed actively during the past
year would severely alter the interpretation of our results.
It was suggested by Pontecorvo that if the neutrino had a
mass it could oscillate between two or more neutrino states,
for example, between neutrinos of the electron type (ν_e) and
neutrinos of the muon type (ν_μ), $\nu_e \rightleftarrows \nu_\mu$. If this occurs
with an oscillation time short compared to the sun-earth
transit time of 500 seconds, the signal would be reduced by
a factor of two or more. Since neutrino oscillations are
suggested by some of the new weak interaction theories,
experimentalists have been stimulated to search for oscillations. Although this may be the solution to the problem of
low observable neutrino fluxes from the sun, it will be
exceedingly difficult to demonstrate this with terrestrially
based experiments.

The calculated neutrino capture cross-sections give the
basic detection sensitivity for the ^{37}Cl detector. We have
proposed to check these calculations experimentally by measuring the cross-section using energetic neutrinos produced
ty muon decay in the beam stop of the Los Alamos Meson Production Facility. Background studies show that this experiment is marginally feasible if the accelerator operates at
its highest beam intensity. However, it now appears unlikely that this measurement will be carried out. Alvarez has
suggested another possibility for measuring the cross-
section--preparing a mega-curie source of ^{65}Zn, in a nuclear
reactor and placing it in or adjacent to the 100,000 gallon
detector. Since this experiment is difficult and expensive
to perform, we do not presently plan to perform it.

Argon Chemistry

There has been the suggestion that an ^{37}Ar atom produced
in perchloroethylene by neutrino capture forms a compound or
molecular association that is not removed by the helium
purge (15). From a chemical point of view it is almost incon-
ceivable that such an entity could exist long enough to influence the experimental recovery of ^{37}Ar. However, in view of
the importance of our results to stellar structure and neutrino physics we did undertake a direct test of this possibility. The test is based on a process that is essentially
identical with that of neutrino capture and electron emission.
This process is the beta decay of chlorine-36
($^{36}Cl \rightarrow ^{36}Ar + e^- + \bar{\nu}$, $t_{1/2} = 3 \times 10^5 y$, beta decay energy 0.71
MeV) in which a low energy antineutrino and an electron are
simultaneously emitted. The recoil dynamics and chemical
fate of the resulting ^{36}Ar atom are essentially identical to

those of an ^{37}Ar atom produced by neutrino capture. Perchloroethylene containing radioactive ^{36}Cl, was synthesized. Seven
millicuries of ^{36}Cl in this chemical form generate 8 x 10^{-7}
cm^3 STP of ^{36}Ar per day. This amount of ^{36}Ar can be readily
measured by neutron activation. The experiment is being
carried out in a 30 liter tank using a helium purge to remove
^{36}Ar. Although this experiment is still in progress (with
Hernán Vera Ruiz), we already know that the yield of ^{36}Ar
removed by a helium purge is essentially quantitative.

The Prospects for
New Solar Neutrino Experiments

The Brookhaven experiment has not been able to observe a
neutrino flux from the sun. It is not clear whether the difficulty lies in a lack of understanding of the theory of solar
structure, of the details of the nuclear reaction processes,
or of the physics of the neutrino. The chlorine experiment
has the disadvantage of having greatest sensitivity to the
flux of energetic neutrinos from ^{8}B decay, and a relatively
low sensitivity to the neutrinos from ^{7}Be and the PeP
reaction. Its energy threshold is too high to observe the
abundant neutrinos from the H-H reaction. The flux of neutrinos from the H-H reaction, or its electron capture branch,
the PeP reaction, are nearly independent of variations in the
solar model. One can argue that if the sun is producing
energy by fusion reactions, it must be using hydrogen as a
primary fuel, and this almost guarantees a flux of H-H
reaction neutrinos at the earth of 6 x 10^{10} cm^{-2} sec^{-1}. From
this viewpoint the most reliable way of testing whether the
sun is radiating observable neutrinos is to measure either the
low energy neutrinos from the H-H reaction or those from the
PeP reaction. If these neutrinos were indeed observed with
the calculated flux one would be assured that the sun is
producing energy by hydrogen fusion, and that the electron-
neutrinos do not decay or oscillate to other states during
their transit to the earth. Such a test of solar energy production processes would have enormous importance in astrophysics and elementary particle physics. However, building a
solar neutrino detector with this capability is not an easy
task.

One possibility would be to build a larger ^{37}Cl experiment with sufficient sensitivity to observe PeP neutrinos.
Such an experiment could also check whether the slightly
positive ^{37}Ar production rate now being observed is indeed
from solar neutrinos. For this purpose one would need a ^{37}Cl
experiment about five times larger than the present one.

Table III

Proposed Solar Neutrino Detectors

Neutrino Capture Reaction	Half-life of Product	Threshold Energy, MeV	Tons of Element needed for 1 ν-capture/day	
			All Sources	H–H or PeP only
$\nu + {}^{71}Ga \to {}^{71}Ge + e^-$	11 days	0.233	38	53
$\nu + {}^{7}Le \to {}^{7}Be + e^-$	53 days	0.862	5	16
$\nu + {}^{55}Mn \to {}^{55}Fe + e^-$	2.6 years	0.231	290	420
$\nu + {}^{205}Tl \to e^- + {}^{205m}Pb \to {}^{205}Pb$	1.6×10^7 years	0.048	13	15
$\nu + {}^{81}Br \to e^- + {}^{81m}Kr \to {}^{81}Kr$	2.1×10^5 years	0.490	660	6900
$\nu + {}^{115}In \to e^- + {}^{115m}Sn \to {}^{115}Sn$	Stable	0.128	3.1	3.3
$\nu + {}^{37}Cl \to {}^{37}Ar + e^-$ (present system)	35 days	0.814	480	10700

Another alternative is to choose a completely new radio-chemical system. During the last few years at Brookaven we have been working to develop a new solar neutrino detector based on a different neutrino capture reaction. A detector with a different energy response will give us more informa-tion about the neutrino energy spectrum. Listed in Table III are the neutrino capture reactions that have been selected after a careful search considering many factors: neutrino capture cross-section, threshold energy, half-life and decay characteristics of the product, availability of the target element, and various background effects (16).

The $^7Li(\nu,e^-)^7Be$ reaction is attractive even though it has a higher energy threshold (0.862 MeV) than the ^{37}Cl re-action (0.814 MeV). Since both nuclei, ^7Li and ^7Be, have the same nuclear structure, they are called mirror nuclei and the neutrino capture reaction has a very high cross section (a super-allowed transition), in fact, the highest possible one. This reaction is therefore an excellent choice for observing mono-energetic (1.44 MeV) PeP neutrinos. The target material, lithium, is relatively cheap, only 15 tons are needed, and the chemical extraction of ^7Be from large amounts of concentrated lithium chloride solution can be done readily. The main difficulty with this system lies in measuring the small number of radioactive ^7Be atoms produced in a solar neutrino experi-ment. At Brookhaven a small pilot experiment is in operation that uses 0.16 toms of lithium in the form of a 12 molar aqueous solution of lithium chloride. Before enlarging this experiment to full scale, it is necessary to develop a sensi-tive method of measuring ^7Be, and to measure the rates of various background processes.

The neutrino capture reaction, $^{71}Ga(\nu,e^-)^{71}Ge$, is by far the best one for observing neutrinos from the H-H reaction. The threshold energy, 0.235 MeV, is low, the neutrino capture cross-section is favorable, and the isolation and counting of ^{71}Ge radioactivity are readily accomplished. Obtaining the 30 to 50 toms of gallium needed for the experiment is the major difficulty. Only recently have ton quantities of galli-um been produced industrially. This element is usually prepared as the metal from the bauxite ores used in the pro-duction of aluminum. It is used to make light emitting diodes for the readout displays of pocket calculators and watches. Since we hope that the quantities needed for a solar neutrino experiment will soon become available, we have built small scale experiments to test recovery procedures and counting techniques. In the near future we will build a pilot system using 200 kg of gallium metal.

Finally, we would like to mention three other techniques
that have been proposed to measure the solar neutrino flux;
two are radiochemical methods, and the third is a direct
counting method. Freedman and his associates at the Argonne
National Laboratory propose using the reaction:

$$^{205}Tl(\nu,e^-) \, ^{205m}Pb \longrightarrow \, ^{205}Pb.$$

Neutrino capture feeds the isomeric state in ^{205}Pb which
rapidly decays to the long lived isotope

$$^{205}Pb \; (t_{\frac{1}{2}} = 1.6 \times 10^7 \text{ years}).$$

Because the energy threshold for this reaction is only 0.046
MeV, it would be sensitive to the H-H neutrinos. By using a
thallium mineral that has been exposed for a long period of
time underground, they hope to test for long period variations
in the solar HH reaction. Performing this experiment requires
locating 3 to 10 kg of the proper mineral, careful chemical
extraction of ^{205}Pb and, finally, measurement of the number
of ^{205}Pb atoms present. Another radiochemical detector was
proposed by Scott of the Scottish Universities Research and
Reactor Centre (also, independently, by Hampel and Kirsten of
the Max-Planck-Institut, Heidelberg). They suggest using the
neutrino capture in ^{81}Br which produces an isomeric state,
^{81m}Kr, that decays to the long lived

$$^{81}Kr \; (t_{\frac{1}{2}} = 2.1 \times 10^5 \text{ years}), \; ^{81}Br(\nu,e^-) \, ^{81m}Kr \longrightarrow \, ^{81}Kr.$$

The main advantage of this reaction is that the product is a
rare gas that can be removed by a helium purge. The capture
reaction has an energy threshold of 0.490 MeV and would observe predominantly the neutrinos from ^7Be decay. The proposed experiment would use a geological deposit as a source
of bromine. However, one could also expose a pure bromine-
containing liquid for a year or more (capture rate = 0.5
atoms ^{81}Kr per year per ton $CHBr_3$) and then remove the ^{81}Kr
that was formed. This approach requires the capability of
observation of a small number of ^{81}Kr atoms using some special
technique, possibly high sensitivity mass spectrometry or
laser-induced fluorescence.

Raghavan of the Bell Laboratories has pointed out that
neutrino capture in ^{115}In will produce an excited state in
^{115}Sn that will rapidly decay (3.2 μsec) emitting two successive gamma rays. This reaction has the advantage of a unique
signal characterizing a neutrino capture event. To build an
indium neutrino detector, one could incorporate indium inside
a scintillation counter that can record this characteristic
event and sort it out from other nuclear processes. He proposed using several tons of indium divided into a modular
array of scintillation counters. Many other techniques could
possibly be used. The important new thought is that this

process has a favorable cross section, a low threshold (0.128 MeV), and a unique signature.

It is clear from the foregoing discussion that there is no easy way to measure the solar neutrino flux. All of the suggested techniques except perhaps the ^{71}Ga-^{71}Ge method require an enormous development to show that they are feasible. Because of the difficulties very few laboratories are involved in a major way with this experimental problem. Outside of our group at Brookhaven and that of our associate, Professor K. Lande of the University of Pennsylvania, there are only two other laboratories with a major effort in neutrino astronomy. One is the group of Professor Fred Reines at the University of California, Irvine, and the other is the group of Professor G. Zatsepin of the Institute for Nuclear Research, Moscow, USSR. As evidenced by the number of papers on the problem, there is great theoretical interest in the field and we hope more experimentalists with fresh ideas will join us in the future.

Research performed at Brookhaven National Laboratory under contract with the U.S. Energy Research and Development Administration.

References

1. H. Bethe, Phys. Rev. 55, 434 (1939); C. F. van Weizsäcker, Physikalische Zeitschrift 38, 176 (1937); I. Iben, Jr., Ann. Rev. Astronomy and Astrophysics 5, 571 (1967), Scientific American, July 1967, p. 27; D. D. Clayton, Stellar Evolution and Nucleosynthesis, McGraw-Hill, 1968; M. E. Burbidge, G. R. Burbidge, W. A. Fowler, and F. Hoyle, Rev. Mod. Phys. 29, 547 (1957).

2. B. Pontecorvo, Chalk River Report PD-205 (1948); L. W. Alvarez, UCRL-328 (1948); See also report on Solar Neutrino Conference, Univ. Calif. at Irvine, F. Reines and V. Trimble, Editors (1973) appendix; V. Trimble and F. Reines, Rev. Mod. Phys. 45, 1 (1973).

3. J. N. Bahcall, Phys. Rev. 135, B137 (1964), Phys. Rev. Letters 17, 398, (1966); G. V. Domogatsky, V. N Gavrin, and R. A. Eramzhyan, Proc. 9th Int. Conf. Cosmic Rays (London) 2, 1034 (1965); W. A. Lanford and B. H. Wildenthal, Phys. Rev. Letters 29, 606 (1972); P. D. Parker, A. J. Howard, and D. R. Goosman, Nucl. Phys. A250, 309, (1975); W. C. Haxton and T. W. Donnelly, Phys. Letters 66B, 123, (1977).

4. Report on the Brookhaven Solar Neutrino Experiment, R.

Davis Jr. and J. C. Evans, BNL-21837. This report contains detailed counting data for the experimental runs reported, and references to earlier reports.

5.   J. N. Bahcall and R. Sears, Ann. Rev. Astronomy and Astrophysics 10, 25 (1972); J. N. Bahcall and R. Davis Jr., Science 191, 264 (1976).

6.   Nuclear reaction rates and opacity
     Thermonuclear Reaction Rates, W. A. Fowler, G. R. Caughlan, and B. A. Zimmerman, Ann. Rev. Astronomy and Astrophysics 5, 525 (1967), 13, 69 (1975); Stellar Opacity, T. R. Carson, ibid. 14, 95 (1976).

7.   The standard solar model
     J. N. Bahcall, High Energy Physics and Nuclear Structure, Proc. 2nd Int. Conf., Rehovoth, G. Alexander, Ed. (North Holland, 1967), p. 232; I. Ivan, Jr., Annals of Physics, N.Y. 54, 164 (1969); J. N. Bahcall, W. F. Huebner, N. H. Magee, Jr., A. L. Merts, and R. K. Ulrich, Astrophys. J. 184, 1 (1973); T. R. Carson, D. Ezer, and R. Stothers, Astrophys. J. 194, 743 (1974); C. A. Rouse, Astron. and Astrophys. 44, 237 (1975).

8.   Reviews
     J. N. Bahcall and R. Sears, Ann. Rev. Astronomy and Astrophysics 10, 25 (1972); B. Kuchowicz, Report on Progress in Physics 39, 291 (1976).

9.   Low heavy element abundances
     J. N. Bahcall and R. K. Ulrich, Astrophys. J. 170, 593 (1971); P. C. Joss, Astrophys. J. 191, 771 (1974); M. J. Newman and R. J. Talbot, Jr., Nature 262, 559 (1976); J. R. Auman and W. H. McCrea, Nature 262, 560 (1976).

10. Periodic sun
     W. A. Fowler, Nature 238, 4 (1972); A. G. W. Cameron, Rev. Geophys. and Space Phys. 11, 505 (1973); F. W. W. Dilke and D. O. Gough, Nature 240, 262 (1972); R. T. Rood, Nature 240, 178 (1972).

11. Stellar rotation and magnetic fields
     P. Demarque, J. Mengel, and A. Sweigart, Astrophys. J. 183, 997 (1973), Nature phys. sci. 246, 33 (1973); R. T. Rood and R. K. Ulrich, Nature 252, 366 (1974); D. Bartenwerfer, Astron. and Astrophys. 25 455 (1973); J. C. Wheeler and A. G. W. Cameron, Astrophys. J. 196, 601 (1975); R. L. Snell, J. C. Wheeler, and J. R. Wilson, Astrophysical Letters 17, 157 (1976).

12. Solar luminosity and the terrestrial temperature
    C. Sagan and A. T. Young, Nature 243, 459 (1973); C.
    Sagan and G. Mullen, Science 177, 52 (1973); R. K.
    Ulrich, Science 190, 619 (1975); R. T. Rood and M.
    Newman, preprint, Cal. Inst. Tech., 1977; L. P. Knauth
    and S. Epstein, Geochim. Cosmochim. Acta 40, 1095 (1976);
    D. L. Turcotte, J. L. Cisne, and J. C. Nordman, Icarus
    30, 254 (1977).

13. Special solar models
    F. Hoyle, Astrophys. J. 197, L127 (1975); J. C. Wheeler
    and A. G. W. Cameron, Astrophys. J. 196, 601 (1975); A.
    J. R. Prentice, Mon. Not. Roy. Astron. Soc. 163, 331
    (1973); R. Stothers and D. Ezer, Astrophys. Letters 13,
    45 (1973); D. D. Clayton, M. J. Newman, R. J. Talbert,
    Jr., Astrophys. J. 201, 489 (1975); C. W. Chin and R.
    Stothers 254, 206 (1975).

14. Neutrino properties
    V. Gribov and B. Pontecorvo, Phys. Letters 28B, 493
    (1969); J. N. Bahcall and S. C. Frautschi, ibid., 29B,
    623, (1969); J. N. Bahcall N. Cabbibo, and A. Yahil,
    Phys. Rev. Letters 28, 316 (1972); S. Pakvassa and K.
    Tennakone, Phys. Rev. Letters 28, 1415 (1972); A. K.
    Mann and H. Primakoff, Phys. Rev. D 15, 655 (1977); S.
    M. Bilenky, S. T. Petcov, and B. Pontecorvo, Phys.
    Letters 67B, 309 (1977); H. Fritzsch and P. Minkowski,
    Phys. Letters 62B, 72 (1976).

15. Argon chemistry
    K. C. Jacobs, Nature 256, 560 (1975); J. J. Leventhal
    and L. Friedman Phys. Rev. D 6, 3338 (1972); B. Banerjee,
    S. M. Chitre, P. O. Divakaran, and D. S. V. Santhanam,
    preprint, Tata Inst. 1977.

16. New solar neutrino detectors
    V. A. Kuzmin, JETP 49, 1532 (1965); J. C. Evans, Solar
    Neutrino Conf., Univ. of California at Irvine, F. Reines
    and V. Trimble, Editors (1973); J. N. Bahcall, Phys.
    Rev. Letters 23, 251 (1969); J. K. Rowley, BNL report
    20397; M. S. Freedman, et al., Science 193, 1117 (1976);
    R. D. Scott, Nature 264, 729 (1976); R. S. Raghavan,
    Phys. Rev. Letters 37, 259 (1976).

Streams, Sectors, and Solar Magnetism

Arthur J. Hundhausen

This paper was originally published as Chapter 7 in the volume Coronal Holes and High Speed Wind Streams, a monograph describing results of the Skylab Solar Workshop on Coronal Holes. Its republication has provided the author with a rare (and welcome) opportunity to correct a few more mistakes, reword some awkward phrases, and add several references that were overlooked in the original writing or that have come to his attention since that time (February, 1977). Despite strong temptations, no attempt has been made to heavily revise or update the article in the light of an additional year of perspective.

Allusions to other chapters of the Skylab monograph have been revised to standard reference form; these individual references (1-6) are, however, listed separately at the head of the bibliography in hopes of directing the reader's attention to other aspects and achievements of the Coronal Hole Workshop. In separating this chapter from its original context, I would like to emphasize several important credits. The two tables in this paper were formulated during the workshop in conjunction with Table 1 of Bohlin (1) and were first printed in a newsletter circulated to workshop participants. Credit for its contents should go to L. F. Burlaga, W. C. Feldman, R. R. Fisher, R. T. Hansen, J. T. Nolte, N. R. Sheeley, Jr., D. G. Sime, J. H. Underwood, and W. J. Wagner.

1. Introduction and Background

The tendency for geomagnetic disturbances to recur with the 27-day synodic rotation period (i.e., that viewed from the orbiting earth) of the equatorial region of the sun was recognized in the mid-nineteenth century (e.g.,(7)) and became important evidence in early arguments for the solar

Reprinted by permission of the Colorado Associated University Press.

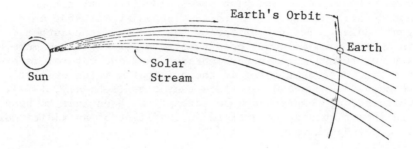

Figure 1:   The geometry of a localized stream
of plasma emitted radially from a steady source
region rotating with the sun (Chapman and
Bartels, 1940).   Solar rotation draws the stream
into a spiral spatial configuration.

control of geomagnetic variations. By the 1920's (see 8 and
9 for further discussion of this and other historical developments), this phenomenon was interpreted in terms of long-
lived streams of ionized solar material, or plasma, emitted
from some localized source fixed on the rotating sun, as
sketched in Fig. 1. The tendency for 27-day geomagnetic recurrence, and hence the implied existence of such streams, was
most clearly demonstrable in the declining phase of each 11-
year sunspot cycle.

The nature of the plasma-emitting solar regions, given
the name M-regions by Bartels (9), remained one of the intriguing mysteries of solar-terrestrial physics for nearly half
a century. The search for this solar cause of a terrestrial
effect was carried out by attempting to demonstrate a valid
and high statistical correlation between a suspected solar
feature and the assumed geomagnetic effect. In the absence
of any observations or understanding of the propagation of a
"signal" through the intervening space, such studies usually
proved to be ambiguous. A fine example of this inherent
difficulty can be found in studies of the relationship of cal-
cium plages (or long-lived centers of weak solar activity) to
recurrent geomagnetic disturbances. Superposed epoch studies
of the geomagnetic response to central meridian passage of
low-latitude plages revealed both a minimum and a maximum in
geomagnetic activity occurring $\sim$ 4 and 6 days after meridian
passage (e.g., 10). In the absence of any information as to
the time required for the solar particle stream to travel
from sun to earth, this result led to schools with diametrically opposed views as to the role of plages in solar particle
emission (see 10-18).

In situ observations of interplanetary space became a
reality in the late 1950's and led to a rapid growth in our
understanding of this region. The existence of a solar wind,
or a continuous rather than intermittent emission of plasma
from the sun, was predicted by Parker (19) and established by
observations made on the Mariner 2 spacecraft enroute to Venus
in 1962. Fortunately, this same mission came during the
declining phase of solar cycle 19 (maximum had occurred in
$\sim$ 1959, with the cycle ending in October 1964) when a sequence
of recurring geomagnetic disturbances was present. The disturbances were found to be associated with solar wind of
abnormally high speed, or with high-speed streams within the
omnipresent flow of solar plasma (20). These streams displayed a recurrence tendency not unlike that of the associated
geomagnetic disturbances. The solar wind speed within a
Mariner 2 stream generally rose from $\sim$ 300 km sec^{-1} to $\sim$ 600
km sec^{-1} within 1 to 2 days and then declined toward the initial value over a longer, $\sim$ 5 day, interval (21).

The subsequent 15 years of in situ interplanetary observations have revealed similar high-speed streams to be common
features of the solar wind. For the epoch 1964 to 1973, the
streams were typically of smaller amplitude (with speeds
ranging from 300 to $\sim$ 500 km sec^{-1}) and less recurrent than
those observed by Mariner 2 (e.g., 22). Other solar wind
properties (the plasma density, proton temperature, direction
of flow, magnetic field intensity, and even chemical composition) have been found to vary systematically through high-
speed streams. Some of this variability may be produced by
varying coronal boundary conditions; however, it has long been
expected and recently demonstrated that the fluid interaction
of high and low speed flows emanating from the corona is in
itself adequate to produce much of the observed variation in
the other solar wind properties mentioned above. This interplanetary evolution is thus a source of confusion in attempts
to associate solar wind and coronal properties.

There is, however, one important characteristic of solar
wind stream structure not subject to this confusion. If averaged over sufficiently long time scales, the interplanetary
magnetic field is oriented in a simple (and presumably well
known) spiral pattern expected for a field frozen into a plasma
expanding from a rotating source (19). The sense of the vector
field, or magnetic polarity, within this basic pattern is
found to be highly organized, with the field pointing predominantly inward (or outward) for many days and then changing on
a shorter time scale to the opposite sense (see the review by
Wilcox, (23)). These often recurrent magnetic sectors are
closely related to high speed streams -- each stream tends to
have a single dominant magnetic polarity with changes in the
large scale polarity generally occurring at low speeds, between the high speed streams. As this sector pattern is
expected to be frozen into the flow, it has served as an important clue in searches for the solar origin of high-speed
streams.

The general understanding of an interplanetary stream-
sector structure outlined above led to some clarification of
the nature of M-regions. Cross-correlation of the observed
interplanetary and photospheric magnetic fields indicated a
relationship with a sun-earth transit time of $\sim$ 4 days; the
interplanetary sectors were apparently related to photospheric
regions of weak fields and a similar dominant polarity (24,25).
Extrapolation to the sun of the observed interplanetary stream
structure under the questionable assumption of constant expansion speed (giving, however, a typical 4 day transit time consistent with that implied by the magnetic cross-correlation)
led to inferred source longitudes unrelated to active centers

(26). Rather, these longitudes generally matched the centers
of large-scale, dominant polarity regions in the photospheric
magnetic field (27). All of these studies were consistent
with the relationship between the interplanetary stream-
sector structure and the large-scale coronal magnetic structure sketched in Fig. 2. This phenomenological description
was also consistent with the modulation of solar wind speed
predicted by simple theoretical models of the coronal expansion in such a magnetic geometry (e.g., 29).

The growing interest in observations of coronal holes in
X-ray and XUV radiation and increasing understanding of their
basic physical characteristics -- low coronal density (30-33)
and open-diverging magnetic fields -- were followed by studies
of the relationship of holes to the solar wind. A correlation
between holes, geomagnetic activity, and high-speed streams
has been reported in many studies (34-37). The implied identification of coronal holes as M-regions is, of course, consistent with the descriptive model of coronal and interplanetary structure shown in Fig. 2 if coronal holes are identified
with open magnetic regions.

2. Coronal Holes, Solar Wind Stream, and Interplanetary Sectors During the Skylab Epoch

The initial motivation for conducting the Skylab Solar
Workshop on Coronal Holes stemmed from an abundance of relevant data. The solar experiment cluster flown on Skylab
itself yielded solar disc observations in X-rays and the
ultra-violet and white-light coronal observations beyond ∿
1.5 solar radii unprecedented in extent and continuity (see
(6) for a description of instruments, institutions, and the
observations). Data obtained from the ground and from other
spacecraft complement the Skylab observations in the visible
solar atmosphere and are the primary source of information on
the interplanetary structure of interest here. The comple-
mentary data also serve to extend our understanding of these
coronal and interplanetary phenomena over a longer time interval and thus place the Skylab epoch in a broader context. It
is in this context that a further motivation for these studies
arises. During the Skylab mission coronal holes, solar wind
streams, and interplanetary sectors were all found to be
evolving in a manner (to be described below) that yields further insight into their inter-relationships. At the very end
of the mission, the corona and solar wind attained a very
simple structure that then persisted, with little further
change, for some two years, and produced a clear long-lived
sequence of recurrent geomagnetic disturbances. Thus the
Skylab mission could hardly have been better timed for study
of the M-region problem, as it witnessed the formation of the

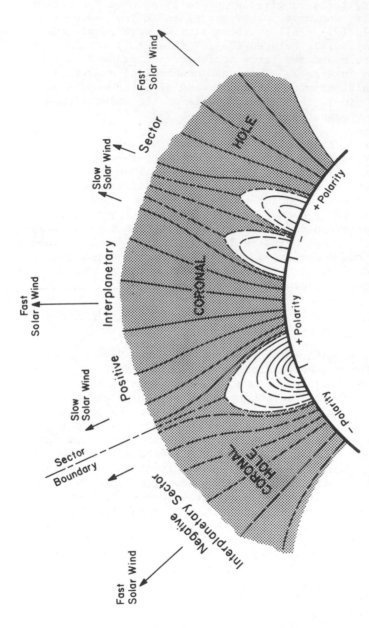

Figure 2. A phenomenological model of large scale coronal magnetic structure and the associated magnetic sectors and high speed streams in interplanetary space (adopted from Hundhausen, 1972, which was modified from earlier drawings by Billings and Roberts, 1964, and Wilcox, 1968). A hypothetical density structure, in which open magnetic regions are identified as coronal holes, has been added to the figure.

classical M-region phenomenon in the declining phase of
solar cycle 20.

a.   Coronal and Solar Wind Conditions, 1972-1976

The evolution of coronal holes during Skylab is described
in some detail in (3) and (4); attention here will be confined
to several features of that evolution crucial to this discussion. Some six coronal holes (the precise number is largely
a matter of definition) have been identified by most observers
during the May 1973 to February 1974 interval of Skylab data
aquisition. The temporal development of individual holes
tended to display the following sequence:
(1)   appearance of a small, isolated hole at low solar lati-
      tudes.
(2)   growth of the area of the feature.
(3)   "connection" to the polar region of the same magnetic
      polarity (positive or outward pointing for the north
      polar region, negative or inward pointing for the south
      polar region).
(4)   shrinking of area of the hole leading ultimately to its
      disappearance.
The entire "birth-to-death" sequence appeared to take place
with little drift in Carrington longitude, requiring 6 to 8
solar rotations in the early part of the epoch.

Of special significance in the present context is the
tendency for the holes to occur in a regular pattern in solar
(Carrington) longitude, spaced $\sim 90°$ (38,39) to $120°$ (40)
apart. This pattern also includes a regular temporal development in which a new coronal hole of a given magnetic polarity
appears and undergoes its development to the east of an older
or earlier hole of the same polarity. These tendencies are
illustrated in Fig. 3 by the locations and dates of existence
of the prominent coronal holes for the Skylab epoch. The
three or four holes generally agreed to be associated with
photospheric regions with a dominant positive polarity (and
all of which "connected" with the northern or positive polarity polar cap at some time during their lifetime) shifted systematically eastward through $\sim 240°$ of longitude during the
~ 10 month Skylab mission; other data (43,40) indicate that
this pattern began in mid-1972 with another hole $\sim 120°$ west
of the first member of the Skylab sequence. Thus these holes
moved eastward in four distinct steps, through $360°$ of Carrington longitude, between mid-1972 and early 1974. Ground-
based observations from 1974 and 1975 (44,40) indicate that
this evolution ceased with the establishment of a simple, two-
hole coronal configuration in 1974.

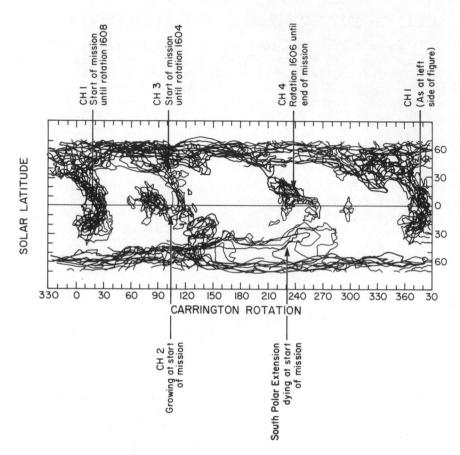

Figure 3:   The locations of important coronal
holes in solar latitude and Carrington longitude
during the Skylab mission.   The positions,
dates, and nomenclature are based on Nolte et
al., 1977, and Bohlin and Rubenstein (1975).
Holes apparently related to photospheric regions
of positive magnetic polarity, or that appeared
to "connect" with the north polar region of the
sun, are labeled from above, while those of
negative polarity or that "connected" with the
south polar region are labeled from below.

Solar wind conditions during the longer 1972-1976 interval can be summarized using published studies or data available at the Coronal Holes Workshop. Daily values of the interplanetary magnetic polarity inferred by Svalgaard (45) from variations in the ground-level geomagnetic field at high latitudes (46) are displayed in Fig. 4, with the time coordinate split into 27-day intervals (or Bartels solar rotations). After mid-1972 a simple two-sector pattern is clearly seen to be the dominant feature in the interplanetary magnetic field. In late 1972 and through most of 1973 this pattern slants across Fig. 4 indicating a recurrence period of about $28\frac{1}{2}$ days; in early 1974 the pattern falls nearly vertically in the figure indicating a recurrence period of about 27 days, or near the synodic (near-equatorial) solar rotation period. The solar wind streams present in 1972 were of normal speed amplitude (i.e., 400 to 600 km sec^{-1}) and duration recurring with a period near 27.1 days (47). Each such stream appeared to persist for 6 to 8 solar rotations. In early 1973 these "normal" streams were joined by a recurrent stream of such heroic amplitude and duration to be classified (affectionately) as a "monster" stream. This feature is illustrated in Fig. 5, a display of the solar wind speed observed by earth-orbiting spacecraft for a 27-day interval in March and April of 1973. The elevation of solar wind speed between April 1 and 6 is typical of solar wind streams observed since 1965; it is dwarfed by the elevation that occurred between March 19 and 28, or in the fully grown monster. In the latter, the solar wind speed remains above 700 km sec^{-1} for some 5 days, or nearly 1/5 of the solar rotation. Such solar wind speeds were previously extremely rare. For example, in 1965-1967 the solar wind speed exceeded 700 km sec^{-1} only 0.2% of the time (48). The monster stream persisted for about 7 solar rotations, producing the largest rise in average solar wind speed since the onset of solar cycle 20 (47). Finally, at the end of 1973, two streams of similar nature (their genealogical relation to the monster has been the subject of debate) developed and then persisted, recurring with a 27.1 day period (47), with little change through 1974 and 1975.

The relationship between coronal holes and the interplanetary sectors and streams described above has been the subject of a number of individual studies (35,36,40,41,43,44). All of these studies are in basic agreement on the following results.
(1) If a large coronal hole exists near the solar equator, a high-speed solar wind stream is observed at the orbit of earth. The magnetic polarity of the stream invariably agrees with the polarity of the coronal hole, as inferred from the dominant polarity of the photospheric region underlying the hole or from the polarity of the polar

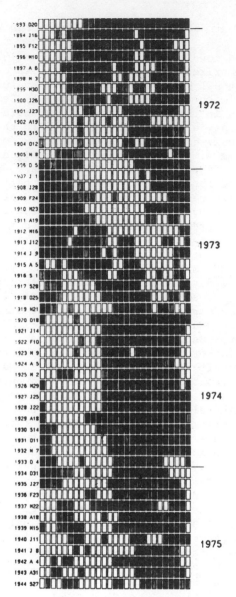

Figure 4: The interplanetary magnetic polarity inferred by
Svalgaard (1976) from high latitude geomagnetic variations for
the years 1972 through 1975. The display is broken into 27-
day Bartels solar rotations, with the rotation number and the
calendar date of the initial day in each rotation indicated
along the left side of the display. The black regions indicate days of predominant negative polarity (or a magnetic
field pointing back toward the sun along the basic spiral
interplanetary field configuration), the white regions indicate days of predominant positive polarity (or a magnetic
field pointing away from the sun), and an intermediate shading
indicates days of mixed or indeterminate polarity.

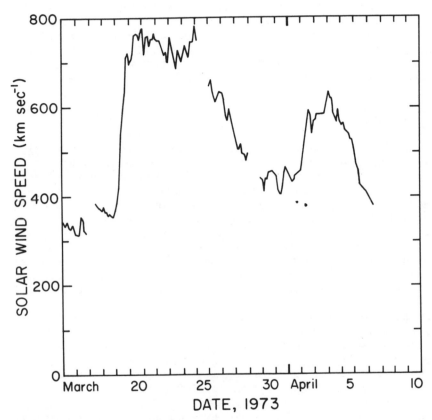

Figure 5: Three hour averages of the solar wind speed observed near the orbit of earth between March 15 and April 10, 1973 (courtesy of S.J. Bame and colleagues, Los Alamos Scientific Laboratory).

region connected to the hole. The amplitude of the solar
wind stream appears to be directly related to the size of
the coronal hole (see 41).

(2) Small coronal holes may not produce high-speed solar wind
streams.

(3) Some solar wind streams occur in the absence of near-
equatorial coronal holes as defined by observations of
the low corona (X-rays, XUV).

At this stage of understanding, essentially that attained at
the beginning of the Coronal Holes Workshop, large, near equatorial coronal holes are a sufficient, but not necessary
condition for the existence of a high-speed solar wind stream.

b. The Polar Regions of the Sun as Sources of Solar Wind Streams

Perhaps the most important contribution of the Coronal
Holes Workshop to our understanding of the relationship of
coronal holes to high-speed streams has been the clarification
of the origin of those recurrent streams occurring in the absence of near equatorial holes. A crucial example of such a
stream is the "monster" described above. Fig. 6 shows the solar wind speed measured on Imp spacecraft at 1 AU and the magnetic polarity of Svalgaard (45) as a function of Carrington
source longitude for Carrington solar rotations 1595-1604, all
in early 1973. The source longitude of any observed parcel of
solar wind has been estimated by the common but controversial
(see 49,50) assumption of radial, constant speed propagation
from the lower corona to 1 AU. The monster stream in its mature form displays solar wind speeds of 700 to 750 km sec^{-1} at
Carrington longitudes between $\sim 180°$ and $\sim 270°$ from rotation
1598 through 1601. Solar wind speeds greater than 700 km sec^{-1}
were first observed in this longitude range on rotation 1597
and were last observed on rotation 1602. The negative magnetic polarity dominant in the monster stream is present at
these longitudes for the entire time covered by this display.

Skylab observations at the estimated Carrington source
longitudes of this stream become available only on rotation
1602 and pertain only to its decay. It is therefore appropriate to examine data covering the entire time interval of Fig.
6. Fig. 7 shows contour maps of the white-light coronal
brightness as a function of solar latitude and Carrington
longitude for solar rotations 1595-1604. These maps are generated from daily ground-based observations of the "polarization brightness" (the difference in white light intensity of
the tangentially and radially polarized components) as a function of solar latitude at 0.5 solar radii above the limb of
the sun; the brightness is then plotted at the Carrington longitude of the limb observation. As the polarization brightness

is essentially proportional to the integral along the line of
sight of electron density, such a map reveals with low resolution (limited by the daily observation cycle and the line-
of-sight integration) the spatial structure of coronal density
if the corona evolves slowly on the solar rotation time scale
necessary to acquire the data set. For the Skylab epoch itself, the low brightness regions in such maps correspond well
to coronal holes observed by other techniques (see (1) and
Fig. 8 below). Assuming that this correspondence remains valid for the time interval of Fig. 7, all regions with a brightness below a fixed low value have been shaded with a pattern
of plus or minus symbols corresponding to the magnetic polarity of the region (again inferred from photospheric observations and the polar connection rule), and will be regarded
henceforth as coronal holes.

A plausible source of the monster stream is apparent in
these maps. A broad equatorward extension of the low brightness or density region associated with the south polar cap of
the sun occupies the approximate 180° to 270° range of Carrington longitude from which the stream is thought to have
emanated. The longitude extent and negative magnetic polarity
of the region match that of the monster stream. Further, comparison of Figs. 6 and 7 reveals a close relationship between
the evolution of the high-speed stream and this south polar
extension. Specifically, the approach of the polar cap extension to the ecliptic plane (shown on Fig. 7 by the heavy line
near the solar equator) between rotations 1596 and 1598 corresponds to the growth of the monster stream; proximity to the
ecliptic (at the 1.5 solar radii heliocentric distance of Fig.
7) in rotations 1598 through 1601 corresponds to the presence
of the fully grown monster, and the non-uniform recession
corresponds to the decay (perhaps splitting in rotation 1602)
and disappearance of the stream. This coronal region was
first suggested as the source of the major geomagnetic disturbances associated with the monster stream by Bell and Nocci
(35). This evolutionary relationship gives strong supporting
evidence for their suggestion.

An immediate implication of this large polar influence
on the solar wind in the ecliptic plane is the appreciable
north-south flow of the expanding coronal plasma. The size
of this effect can be estimated by turning to the Skylab
observations from solar rotation 1602. Fig. 8 combines the
X-ray synoptic map for this rotation with coronal brightness
contour maps based on observations at three heights, 0.2, 0.5
(as before), and 0.8 solar radii, above the photosphere. The
sharp edge of the dark south polar cap hole in the X-ray map
is extended toward the solar equator between $\sim 140^{\circ}$ and 290°,
reaching to within 20° to 30° of the equator (and the nearly

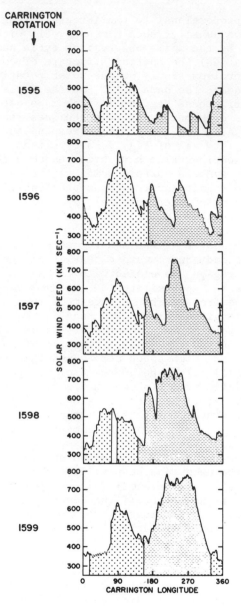

Figure 6:   Observed solar wind speed and magnetic polarity as a function of estimated source
longitude for Carrington solar rotations 1595-
1604, in late 1972 and early 1973.   Three-hour
averages of speed observed by Imp spacecraft
have been carried back to 20 solar radii assuming radial propagation at the observed speed to

CARRINGTON
ROTATION
↓

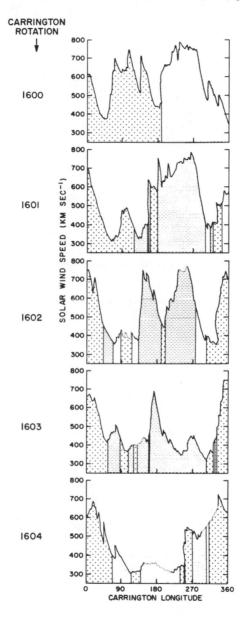

estimate the source longitude; the daily magnetic polarity inferrences of Svalgaard (1976)
have been transformed to longitude in the same
manner and indicated on the figure by plus and
minus signs (courtesy of S.J. Bame and colleagues,
Los Alamos Scientific Laboratory).

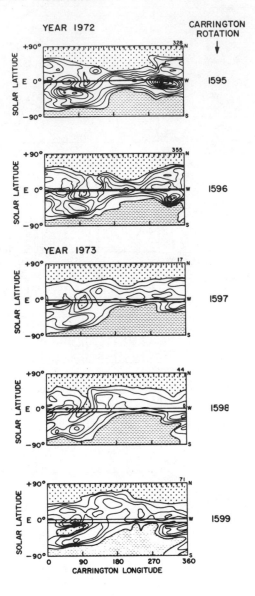

Figure 7:   Contour maps of the coronal brightness as a function of solar latitude and Carrington longitude for Carrington solar rotations
1595-1604. Daily values of the polarization
brightness (see the text), observed as a function of latitude at 0.5 solar radii above the
west limb of the sun by the K-coronameter at the
Mauna Loa Observatory, were used to generate

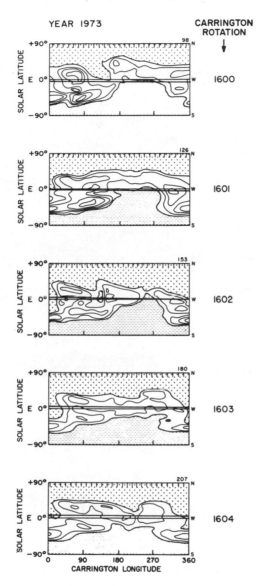

these maps by assigning the observed brightness
to the Carrington longitude of the limb at the
time of observation. All regions with a brightness below a fixed, low level have been shaded
with symbols appropriate to the dominant magnetic
polarity of the underlying photosphere (courtesy
of R.T. and S.F. Hansen, High Altitude Observatory).

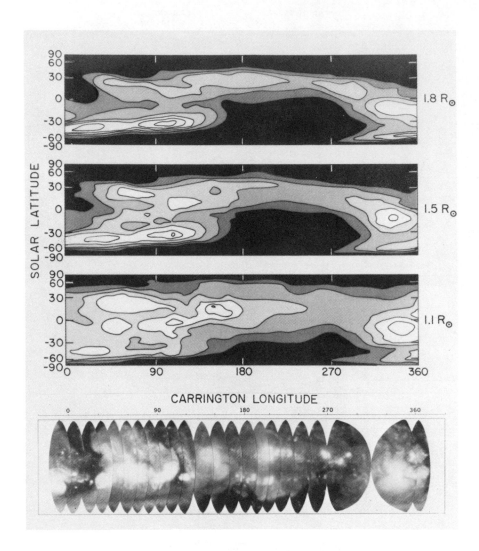

Fig. 8.   The X-ray synoptic map (courtesy of A. S. Krieger and colleagues at American Science and Engineering) and white light coronal brightness maps (courtesy of R. T. and S. F. Hansen, High Altitude Observatory) at three different heights, for Carrington Solar rotation 1602 in June 1973.

coincident ecliptic at this time of year) only at $\sim 180°$ and 270° longitude and lying as far as 50° to 60° south near 200° longitude. The "split" character of the high-speed stream during this rotation suggests origin of the fastest wind in the two regions where the polar cap hole reaches closest to the equator. However, the observation of a predominantly negative magnetic field for all but a small band of longitude in the range 130° to 280° indicates that magnetic field lines may reach as much as 50°-60° northward from their solar foot-points. The changes in white light coronal structure with altitude suggest a similar effect; the low brightness or density region appears to move northward with increasing height, with the bright corona becoming constricted to a narrow band well north of the equator. This latter effect suggests that a coronal streamer, containing a neutral sheet separating the inward pointing open magnetic fields of the south polar region from outward pointing fields emanating from a similar north polar region, might lie above this east-west band of bright corona. Fig. 9a shows the white-light corona photographed by the Skylab coronagraph on June 5, 1973, when the region near 210° Carrington longitude was at the east limb of the sun. A streamer-like structure does indeed rise above the dark occulting disk (at ~ 1.5 solar radii) at $\sim 25°$ north of the equator (see 51). The consistency of this observation with the ground-based coronameter observations of Fig. 8 is emphasized by an overlay of contours of the brightness observed on the same day at Mauna Loa on the Skylab image. If magnetic field lines from the south polar extension do approach this streamer in the outer corona or interplanetary space, they must be displaced from between 20° and 50° south latitude to the streamer latitude of $\sim 25°$ north, or by no less than 45° and possibly as much as 75°. A simple magnetic geometry consistent with the observed coronal and interplanetary structures is sketched in Fig. 9b.

Variations in the observed interplanetary magnetic field associated with the ± 7° annual excursion of the earth in solar latitude have been presented in the past as evidence for a polar dipole influence on the magnetic structure in the ecliptic plane (52-54). Thus the viewpoint advocated above is not unprecedented. However, the size of this influence may be unexpectedly large and the special character of the monster stream is of such interest that additional evidence for its origin in a polar caplike feature would be welcome. Further evidence is found in the coronal and interplanetary structures observed at the very end of and after the Skylab mission. In late 1973, the evolution of coronal holes led to formation of a negative polarity hole near Carrington longitude 120° (Coronal hole 2 in the catalog of Nolte et al. (41)) and a positive polarity hole near longitude 300°

Fig. 9.   A photograph of the white-light corona above the
east limb of the sun on June 5, 1973 (courtesy of
R. M. Mac Queen and colleagues, High Altitude
Observatory) is shown in part (a); the dark sun-
centered semi-circle is the occulting disk of the
instrument, blocking off radiation within ∿ 0.5
solar radius from the limb.   Superposed on the photo
are contours of the white-light coronal brightness

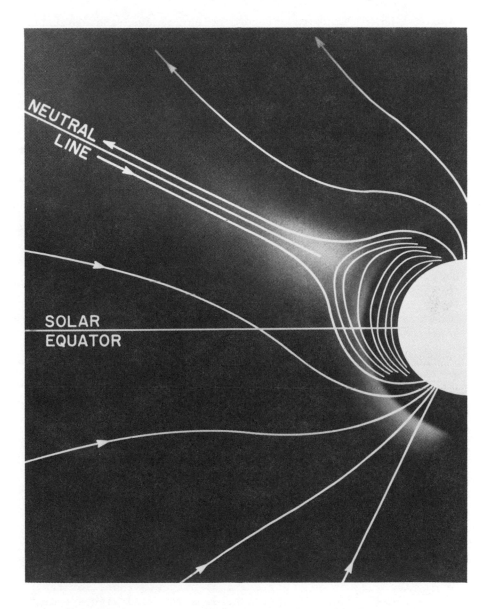

observed above the east limb on the same day by the
K-coronameter at the Mauna Loa Observatory. The
distant coronal streamer in the Skylab photo, suggesting the neutral line extending into interplanetary space, falls directly above the brightest
outermost region observed at lower altitudes. Part
(b) shows a hypothetical open magnetic field geometry
consistent with the observed brightness structure.

(Coronal hole 4). Both of these holes followed the usual
pattern of a real growth and connection to the polar cap of
like magnetic polarity. By the end of the Skylab mission in
solar rotation 1610, the areas of both features had become so
large that they each resembled the polar cap extension advocated
above as the source of the monster stream. Figure 10 shows
the white-light coronal brightness map at 0.5 solar radii
above the photosphere for solar rotation 1616 (as in Fig. 7)
and the solar wind speed and magnetic polarity as a function
of Carrington longitude (as in Fig. 6). High-speed solar wind
streams of the proper polarity again seem to be associated
with these polar cap extensions; the speed within these streams
again reaches the 750 km sec^{-1} level that characterized the
original monster stream. Fig. 11 shows a single X-ray image
of the sun obtained during a rocket flight on June 27, 1974,
or when the north polar extension near 240° Carrington longitude on Fig. 10 was near central meridian. The dark X-ray
hole in this case clearly extends to the solar equator, so
that no large excursion in latitude need be invoked to explain
the association of 750 km sec^{-1} solar wind with this feature.
The presence of two such sources suggests that the neutral
line separating the open magnetic field lines of opposite
sense emanating from the two polar regions lies north of the
equator at Carrington longitudes between 0° and 180° and south
of the equator at longitudes between 180° and 360°. The two-
sector structure of the interplanetary magnetic field is then
readily explained by the rotation of this "tilted dipole-like"
structure (i.e., with different magnetic polarities dominating
opposite hemispheres of the sun, and with those hemispheres
tilted relative to the solar rotation axis). Both this coronal structure and the associated two-sector, two-stream interplanetary pattern persist with little change through 1974 and
1975. At times the general pattern of coronal brightness
suggests a "tilted dipole" to a remarkable degree; the coronal
symmetry axis generally appears to be at an angle of $\sim$ 30° to
the axis of solar rotation (see section 3).

The consistent appearance of exceptionally fast solar
wind in the streams associated with large polar-cap extensions
and the broad, flat-topped nature of some of these streams
(e.g., the monster on Figs. 5 and 6) suggests that, for the
time interval under discussion here, solar wind with speeds
of 700–750 km sec^{-1} always emanated from the polar caps of
the sun, but was observed in the ecliptic plane only when the
polar cap holes extended sufficiently close to the ecliptic.
Strong support for this suggestion stems from observations of
solar wind speed away from the ecliptic plane made using radio
scintillations (55). Figs. 12 and 13 show regions of high and
low solar wind speed as functions of latitude and Carrington
longitude for the first half of 1973 and 1974. Both maps

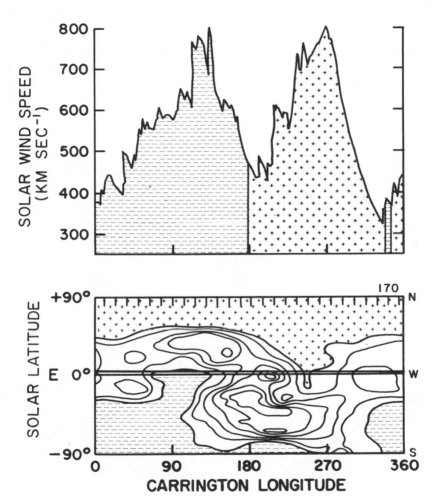

Figure 10:   The coronal brightness contour map
(as in Fig. 7) and observed solar wind speed and
magnetic polarity variations with estimated
Carrington source longitude (as in Fig. 6) for
Carrington solar rotation 1616.

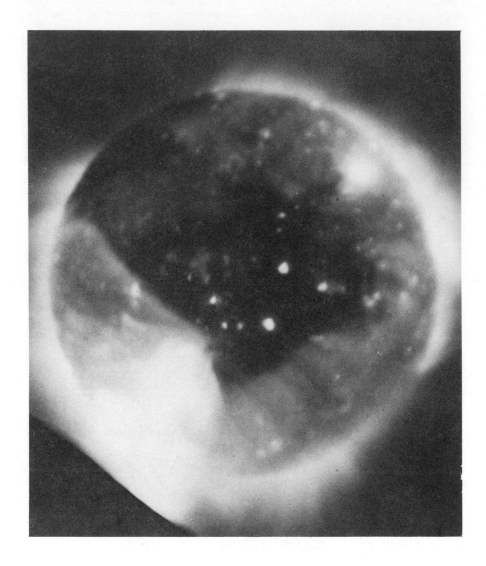

Fig. 11.   An X-ray image of the sun obtained on June 27, 1974
           (courtesy A. S. Krieger and colleagues, American
           Science and Engineering).

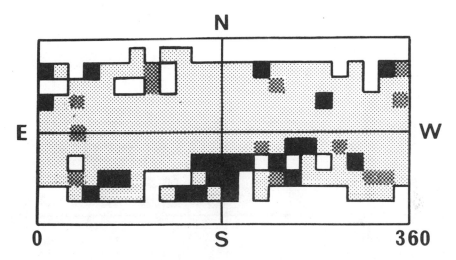

Figure 12:   A map of the solar wind speed as a
function of heliographic latitude and Carrington
longitude, as inferred from radio scintillation
observations made in early 1973 (Carrington
rotations 1599-1604).   Average speeds in 15° x
15° bins are indicated by
(a) light gray shading where the average speed
is below 550 km sec^{-1}
(b) dark gray shading where the average speed is
between 550 and 600 km sec^{-1}
(c) black where the average speed is greater
than 600 km sec^{-1}.
The white areas indicate an absence of observations (Figure courtesy, B. Rickett and D. Sime,
University of California at San Diego; see Sime,
1976).

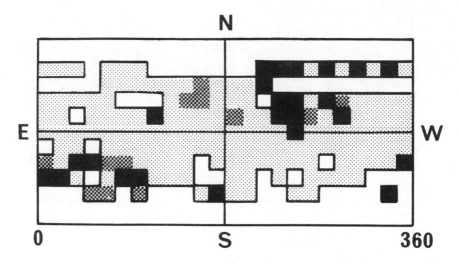

Figure 13: A contour map of the solar wind
speed as a function of latitude and Carrington
longitude as inferred from radio scintillation
observations made in 1974. (Carrington rotations 1612-1616).

show broad regions of high-speed flow at high latitudes, extending into the ecliptic plane at the longitudes where high-
speed streams were observed by in situ probes (Figs. 6 and 10).
These extensions also correspond well to the equatorward extensions of the polar cap. Thus the three dimensional structure
inferred indirectly from observations in the ecliptic is well
confirmed by the available out-of-ecliptic data.

c. The General Relationship Among Holes, Streams and Sectors

Given this evidence that high-speed solar wind emanates
from the magnetic polar caps of the sun and that this high-
speed wind can reach the ecliptic plane if these regions extend sufficiently close to the solar equator (within 20° to
30° for the monster), let us return to the problem of the
relationship between coronal holes, with their definition extended to include such polar cap "extensions", and high-speed
wind streams observed in the ecliptic. In the course of the
Coronal Holes Workshop, a compendium of coronal holes defined
by all observationsal techniques was compiled and the compati-
bility of differing definitions examined (see 1). In Table 1
of that paper all coronal holes in the compendium that extended to within 30° of the solar equator are listed, and solar
wind conditions extrapolated (as in Figs. 6 and 10) to the
same Carrington longitude are briefly described. In addition,
any remaining streams of high-speed wind are noted. Of the
33 appearances (including recurrences) of such coronal holes
or polar extensions for which solar wind observations at the
matching longitude were available, 26 (or 79%) were clearly
associated with high-speed streams of the appropriate magnetic
polarity. Of the 7 remaining holes, two, both small holes,
could be associated with "remnants" of streams--very small-
amplitude streams preceded by large shells of compressed plasma. Such structures would be expected if a stream of short
duration were damped in transit to 1 AU by piling up a large
amount of slower wind (see 56,57). Thus the test of suffi-
ciency for the association of a high-speed solar wind stream
of the correct magnetic polarity with any coronal hole in this
list is satisfied in between 79% and 85% of available cases.
A still higher rate of association is obtained if a more con-
servative list of holes is employed--for example, dark X-ray
regions for which an open magnetic structure is suggested by
structure in the neighboring bright regions (41) or the white-
light brightness maps in which small coronal holes are systematically excluded (as in Fig. 7). For these two sets of coronal holes, the association holds in 90% of the tests. Con-
versely, of the 34 prominent streams of high-speed wind noted
in Table 1 of (1), 27 could be clearly related to coronal
holes of the proper magnetic polarity (a 79% success ratio).
Of the 7 remaining streams, 4 appeared to be associated with

TABLE ONE: MAGNETIC FLUXES IN SOLAR WIND STREAMS AND CORONAL HOLES

EXAMPLE 1: Coronal Hole 1 and stream of June 4 at 1 AU.

Extent at 1 AU: Start at 1200, June 1 (speed = 310 km sec^{-1}) $\Big\}$ 89° at 1 AU.

End at   0600, June 8 (speed = 350 km sec^{-1})

Average radial component of B: B   = 4.4 x 10^{-5} gauss in "flat region" near peak of stream,

June 4.

$$\langle B_r \rangle = 4.5 \times 10^{-5} \text{ averaged over stream.}$$

Ratio of areas of stream and hole: assuming a north-south lane,

$$\frac{A_i}{A_b} \approx \frac{215 \ R_o}{R_o} {}^2 \frac{\text{longitude width of stream}}{\text{longitude width of hole}}$$

$$= (4.3 \times 10^4) \ (5.24)$$

$$\approx 2.4 \times 10^5 \ .$$

Inferred $\langle B_r \rangle$ in hole is then $\langle B_r \rangle = 4.5 \times 10^{-5}(2.4 \times 10^5)$

$$= 10.7 \text{ gauss.}$$

EXAMPLE 2: South Polar Extension and Monster Stream.

Extent at 1 AU: Carrington Rotation   1598   112° for speed elevation
                                          1599   110
                                          1601   101
                                          1602   131

Average radial component of B: Rotation   1598 $\langle B_r \rangle = 2.8 \times 10^{-5}$ gauss
                                          1599 4.2×10^{-5} gauss
                                          1601 3.9×10^{-5} gauss
                                          1602 3.5×10^{-5} gauss

Average of these is $\langle B_r \rangle = 3.6 \times 10^{-5}$ gauss.

Ratio of areas of stream and hole assumed to be $(215)^2$ times the 7 to 1 solid angle ratio of Munro and Jackson (94) for the north polar hole (not the same region), or

$$\frac{A_i}{A_b} \approx 3.2 \times 10^5$$

Inferred $\langle B_r \rangle$ in south polar cap is then $\langle B_r \rangle = 3.6 \times 10^{-5}(3.2 \times 10^5)$

= 11.6 gauss.

EXAMPLE 3: CH2 and Huge Stream on rotation 1607.

Extent at 1 AU: Magnetic polarity changes give sharply defined bounds:

Start at 2100, Nov. 3 ⎱
                     ⎰ 10 days, 9 hours = 137° longitude.
End at 0600, Nov. 13

Average radial component of B is $\langle B_r \rangle$ = 4.2 x 10^{-5} gauss.

Ratio of areas of stream and hole: assuming no large north-south flow,

$$\frac{A_i}{A_b} = (215)^2 \times \frac{\text{longitude width at 1AU}}{\text{longitude width of hole}}$$

$$= (215)^2 (3.1)$$

$$= 1.4 \times 10^5 .$$

Inferred $\langle B_r \rangle$ in hole is then $\langle B_r \rangle$ = (4.2 x 10^{-5}) (1.4 x 10^5)

$$= 6 \text{ gauss.}$$

EXAMPLE 4: CH4 and Stream of Nov. 28 at 1 AU.

Extent at 1 AU: Magnetic polarity changes give sharply defined bounds:

Start at 0600, Nov. 23 ⎞
⎟ 10 days, 18 hrs = 140° longitude.
End at   1200, Dec. 3 ⎠

Average radial component of B: $<B_r> = 3.5 \times 10^{-5}$ gauss.

Ratio of areas of stream and hole: again using the longitude width,

$$\frac{A_i}{A_b} = (215)^2 (\frac{140°}{27°}) = 2.4 \times 10^5 .$$

Inferred $<B_r>$ in hole is then $<B_r> = 3.5 \times 10^{-5} (2.4 \times 10^5)$

$$= 8.3 \text{ gauss.}$$

polar cap extensions falling just outside of the $\pm 30^{\circ}$ latitude
range adopted at the start of this study, while only 3 had
no plausible source in a coronal hole or polar cap extension.
Thus this test of necessity for such features to be present
at every high speed stream is met in 79% to 91% of the cases.

We are thus led to a relationship between coronal holes
and high-speed wind streams that is close to one-to-one for
the Skylab epoch. The most pronounced difficulty encountered
in earlier studies, the observation of some large high-speed
streams not related to equatorial holes, is largely eliminated
by inclusion of the polar cap extensions in the list of
possible coronal sources.

Implicit in the high level of association deduced above
is another piece of evidence for the physical relationship
between holes and high-speed streams, namely their closely
related patterns of evolution. This evolutionary relationship
has been illustrated above for the monster stream and south
polar cap extension in early 1973; it holds equally well for
each coronal hole displaying a slow evolution during the Skylab epoch (40). In general, the growth in area of a low-
latitude hole or the approach of a polar extension toward the
ecliptic plane is accompanied by a rise in the maximum speed
of the associated stream. For moderately large equatorial
holes this relationship is expressed in the hole size versus
stream amplitude relationship of Nolte et al. (41). For very
large holes or polar extensions within $\sim 30^{\circ}$ of the ecliptic,
the maximum solar wind speed approaches the 700-750 km sec^{-1}
level characteristic of the monster stream and the two stream
structure of 1974-1975, with little further change related to
size or position. An interesting extension of this relationship concerns small, young holes and polar cap extensions more
than $\sim 40^{\circ}$ from the ecliptic plane. Although such features
do not produce prominent streams of high speed wind at the
orbit of earth, several studies (e.g., 41) suggest that they
do affect the polarity of the interplanetary magnetic field
in the ecliptic. That is, small coronal holes may produce
solar wind with their own dominant magnetic polarity, but
with low solar wind speeds as observed at 1 AU, and polar
extensions may influence the magnetic polarity in the ecliptic plane even though not sufficiently close to the ecliptic
to impose their characteristic high flow speeds upon it (40,
43).

The related temporal evolution of coronal holes and interplanetary features assumes greater importance when interplanetary magnetic sectors are considered. We have already
noted that during the time interval when coronal holes were
observed to be evolving significantly, or before the beginning

of 1974, the magnetic sectors and high speed streams displayed
a ∿ 1.5 day difference in recurrence periods. Thus the commonly held view that sectors and streams are essentially identical must be an oversimplification when applied to the Skylab
epoch. The tendency for coronal holes to show little drift in
the Carrington longitude system (see 3), and to be closely
associated with high-speed streams whose recurrence period is
extremely close to 27.1 days (47) raises an interesting question with regard to the source of the related magnetic sectors
whose recurrence period is ∿ 28.5 days. This minor paradox is
resolved by the systematic eastward shift in the pattern of
coronal holes of a given magnetic polarity already described
above. That is, while a given hole and high-speed stream of,
say, positive polarity remain fixed in a coordinate frame
with a 27-day period, the growth of a new hole and stream of
the same polarity to the east, and the subsequent decay and
disappearance of the initial feature leads to an eastward
drift of the magnetic polarity pattern carried into interplanetary space by the streams. The eastward drift through 360°
longitude or one full solar rotation in an interval of ∿ 18
months or 20 solar rotations implies a difference in rotation
periods of ∿ 1.4 days. The corresponding recurrence period of
28.4 days for interplanetary sectors agrees well with that
implied by Fig. 4 and inferred more generally for the two-
sector pattern by Svalgaard and Wilcox (58).

All of this evidence points to a strong physical cause
and effect relationship between coronal holes (including polar
cap extensions) and the interplanetary stream-sector structure. There seems to be little justifiable doubt that high-
speed streams originated in coronal holes during the 1972-
1976 time interval, and that the development of the two-hole
corona and two-streams, two-sector interplanetary structure
in 1974 led to the most prominent sequences of recurrent geomagnetic activity in the past solar cycle. Thus the coronal
holes present in 1974, so large as to effectively tilt the
sun's magnetic polar caps, are the most prominent M-regions in
solar cycle 20. The appearance of the monster stream in 1973
and the long-lived, similar pair of streams in 1974-1975 are
the major changes in solar wind conditions observed in the
ecliptic plane during this solar cycle. We are thus confront-
ed with an important new concept of coronal and solar wind
structure to be incorporated into phenomenological and physical models of the expanding corona.

3. The Large-Scale Coronal and Interplanetary Structure Related to Coronal Holes

The observations summarized above clearly reveal some
important features of the three-dimensional, large-scale

structure in the corona and interplanetary space. However, the available data is insufficient to yield a complete description of these regions. For example, observations of the visible layers in the corona pertain most directly to the spatial variations of electron density. Extremely detailed and comprehensive measurements of interplanetary plasma and magnetic field properties are available for the region accessible to in situ probes, near the orbit of earth; at smaller heliocentric distances and outside of the ± 7° band of solar latitude spanned by the ecliptic plane our knowledge (for the Skylab epoch) is limited to expansion speeds and levels of density fluctuations inferred from radio scintillations. Thus some interpretive framework is required to unify this heterogeneous and incomplete data set.

In an ideal scientific world, such a framework would stem from a physical model of the solar corona; for example, solution of the conservation equations of magnetohydrodynamics (including, of course, all important physical processes) for realistic or observed boundary conditions. At present, however, no such model is available; our understanding of coronal physics is far from complete and solution of the mhd equations is mathematically tractable only under severely restrictive simplifying assumptions and extremely simple boundary conditions. Thus we must search for an interpretive framework in a phenomenological model or description, based on assumptions consistent with the available observations and the broad understanding gained from idealized theoretical studies of the expanding coronal and interplanetary plasma. In fact, many of the observations described above can be incorporated into a simple phenomenological description in which the unifying element is the solar magnetic field.

a. Formulation of a Phenomenological Description

A simple basis for the interpretation of many coronal observations is the assumption of a close relationship between density or brightness structure and the large-scale geometry of the magnetic field. This tenet in the folklore of solar physics is not founded on direct evidence, as actual observations of coronal magnetic fields are extremely rare. Rather, it is based on the general appearance of brightness striations in the corona, the comparison of observed structures with theoretical extensions of measured photospheric magnetic field (see 59,60) and the physical expectation that diffusion across magnetic field lines should be small at low coronal densities. In particular, the appearance of coronal holes and the surrounding bright regions in X-rays or ultra-violet radiation strongly suggests (see 3,4) that:
(1) The low density regions identified as coronal holes in

radiation emitted from or scattered by the corona are regions of largely "open" magnetic field lines (i.e., field lines that close only in the distant reaches of interplanetary space) diverging more rapidly than a radial field. Smaller regions of a closed magnetic geometry, as in Fig. 2, may lie within the outer boundaries of a coronal hole.

(2) the neighboring bright or high-density regions consist largely of magnetic field lines that are closed near the sun.

These simple identifications of low coronal density with an open magnetic geometry and high coronal density with a closed magnetic geometry have already been used to superpose a coronal density structure upon the hypothetical magnetic configuration of Fig. 2. Application of the same identifications to an observed three-dimensional representation of coronal brightness or density (such as Fig. 8) yields a qualitative description of the three-dimensional coronal magnetic geometry.

In extending this magnetic geometry to the outer corona and interplanetary region, beyond the observed density structure, we will concentrate on the pattern of magnetic polarities in the open regions. At sufficiently large heliocentric distances, the supersonic and super-Alfvénic outward flow of solar wind must draw the frozen-in field lines into the observed spiral interplanetary field configuration (e.g., 19, 61). Open magnetic regions of opposite polarity must then be separated by neutral surfaces at which the magnetic field reverses (as at the neutral line in the cross-sectional view of Fig. 9b). These neutral surfaces will intersect any sun-centered sphere in one or more neutral lines. We will thus map the coronal magnetic geometry into the interplanetary region by assuming:

(1) that the neutral lines in the high corona (beyond $\sim$ 2 solar radii above the photosphere) lie over the brightest or densest regions of the lower corona.

(2) that the neutral lines map into interplanetary space along spiral field lines, with no other distortion.

The first of these assumptions has already been invoked in drawing Fig. 9b. The second neglects the distortion that must occur in any inhomogeneous solar wind flow. Quantitative models of this process for an axially symmetric coronal expansion (62) and for more general inhomogeneities in the supersonic regime (63) indicate displacements of field lines by less than $\sim$ 10° in latitude or longitude (relative to the spiral geometry) between a few solar radii and the orbit of earth and justify their neglect here.

Any such mapping is subject to an observational constraint; the resulting interplanetary geometry must correspond

to the observed magnetic sector structure. Unfortunately, the
sector structure is directly observed only in the ecliptic
plane, and this constraint is weak. Given a three-dimensional
magnetic polarity map that does agree with the ecliptic plane
sectors, the well-known correlation between solar wind speed
and sectors will now be assumed to hold in three dimensions
and used to infer the variation of speed over the sun-centered
sphere at 1 AU. The discussion of Section 2b suggests that:
(1) the solar wind speed is always about 300 km sec^{-1} at any
     neutral line (or sector boundary).
(2) the solar wind speed attains the "monster stream" level
     of 700-750 km sec^{-1} beyond some angle θ from any neutral
     line.
In other words, we will take the solar wind speed to be that
inferred for flow from the polar caps in 1973-1975 except in
"belts" of angular width 2θ centered on any neutral line. The
angle θ will be evaluated below for several specific magnetic
geometries from the 1973-1975 epoch.

We thus have an _empirical_ scheme that associates a three-
dimensional magnetic geometry with observed corona densities,
extends this geometry into interplanetary space, and uses the
inferred polarity pattern to extend the interplanetary data
set out of the ecliptic plane. Some insight into the plausibility of the scheme can be gained by comparison with the
only known _physical_ model for the magnetohydrodynamic equilibrium state of the solar corona. Pneuman and Kopp (29) and
Endler (64) have obtained numerical solutions to the mhd equations for an isothermal coronal plasma in a solar magnetic
field assumed to be a dipole at the base of the corona. Fig.
14 illustrates the magnetic field lines computed by Pneuman
and Kopp (29) in a plane containing the dipole axis (along the
left margin of the figure). Near the dipole equator the coronal plasma has distended the magnetic field lines (shown by
the dashed lines for the pure dipole) away from the sun; however, these field lines remain closed, containing a coronal
plasma in magnetostatic equilibrium. Near the dipole axis
the outward pressure of the plasma is sufficiently strong to
force the field lines to open, permitting a steady expansion
into interplanetary space. The field lines extending outward
from these open dipolar caps diverge rapidly to fill the space
above the closed field lines. Beyond $\sim$ 1.5 solar radii above
the base of the corona all field lines are drawn outward by
the expanding plasma. A neutral sheet, separating field lines
that originated in opposite "dipolar caps" and hence with
opposite magnetic polarities, occurs at the dipole equator.
For a uniform density at the base of the corona, the density
in the open magnetic regions will fall off more rapidly with
height than on the closed field lines (65). This quantitative
physical model, based on such simplifying assumptions as an

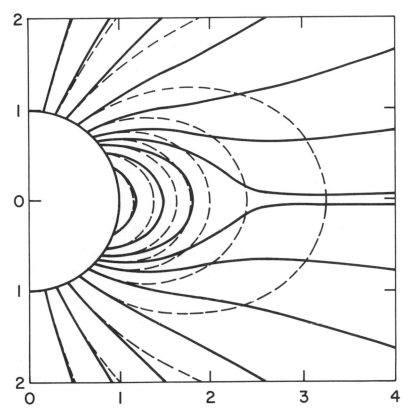

Figure 14: The magnetic field geometry computed by Pneuman and Kopp (1971) for equilibrium of an isothermal corona with a dipole magnetic field imposed on the base of the corona.
Expansion into interplanetary space occurs
along the open magnetic field lines from the
"polar" regions. The dashed lines indicate a
pure dipole geometry.

isothermal corona and with the simplest of solar magnetic boundary conditions, a dipole, leads to a three-dimensional density structure and magnetic geometry (obtained by rotating Fig. 14 about the dipole axis) illustrated in Fig. 15. If the dipole axis is at an angle α to the solar rotation axis, the neutral sheet will be inclined at the same angle to the solar equatorial plane. A map of the coronal density structure would have the general appearance shown in Fig. 16, with a band of dense low corona (the closed field lines) inclined at the angle α to the equator and thus displaying a sinusoidal pattern as a function of latitude and longitude. Given such a coronal brightness map as its starting point, the empirical scheme described above would reproduce the proper qualitative density and magnetic structure, and in particular, a reasonable neutral line geometry well away from the sun. Thus a phenomenological model based on this scheme does correspond with the predictions of a true physical model for this simple magnetic geometry.

b.   Some Specific Applications of the Phenomenological Model

Let us now apply this scheme to coronal data for several solar rotations during and after the Skylab mission. Fig. 17 shows contour maps of the coronal brightness at 0.5 solar radii above the photosphere for Carrington solar rotations 1602 (as included in Fig. 7), 1616 (as in Fig. 10) and 1627. A single magnetic neutral line encircling the sun has been drawn on the map for each rotation. Although ambiguities do exist with regard to the detailed shapes of these lines, their general form is well defined by the observed brightness structure and the constraint that they match the observed interplanetary sector structure.

Carrington rotation 1602 is the most complex of these cases and has already been extensively discussed in Section 2 (see Figs. 8, 9, and 12). The neutral line drawn for this rotation (Fig. 17a) is displaced north of the equator between longitudes 135° and 300°, above the negative polarity south polar cap extension, and south of the equator between $\sim 300^{\circ}$ and 120°, below the two smaller coronal holes with the positive polarity of the north polar cap. Thus the basic magnetic geometry of the distant corona and solar wind bears some resemblance to that expected for a tilted dipole. The region between $\sim 0^{\circ}$ and 90° latitude shows an additional complexity, however, in the coronal brightness structure, in the presence of two coronal holes and high-speed streams, and in the presence of a smaller negative polarity "incursion" in the basic two sector pattern. In other words, the tilted dipole is but one component, albeit a strong component, of the coronal magnetic structure. The three-dimensional structure that might

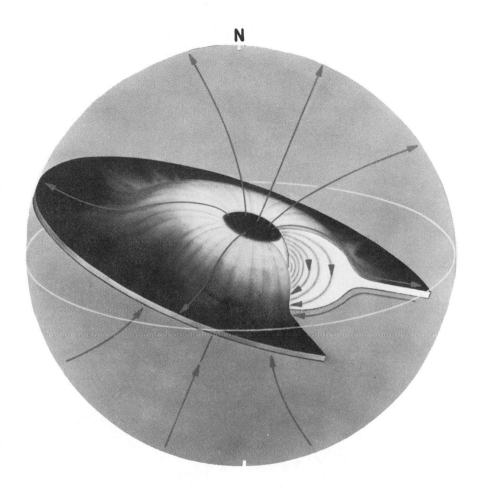

Fig. 15.   An artist's drawing of the three-dimensional
           magnetic and density structure implied by the
           dipole model of Pneuman and Kopp (<u>29</u>). The dense
           (bright) corona on closed magnetic field lines is
           shown as blending into a neutral surface well away
           from the sun. The two open dipolar caps (dark area
           at the base of the corona) are the origins of
           interplanetary magnetic field lines (shown in gray)
           of opposite polarity, separated by the neutral
           surface.

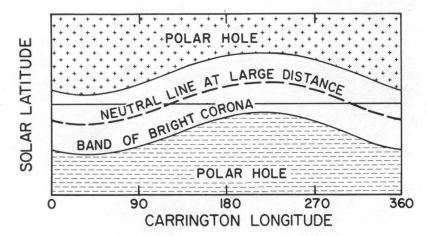

Figure 16: The general form of the coronal
brightness map that would be observed if the
corona had the simple three-dimensional density
structure of Fig. 15. The tilted dipole would
lead to the sinusoidal band of bright corona
shown in the figure, with a neutral line at
large heliocentric distances rising above the
bright structure as shown. A two-sector, two-
stream interplanetary structure should arise
from the intersection of this structure with
the ecliptic plane.

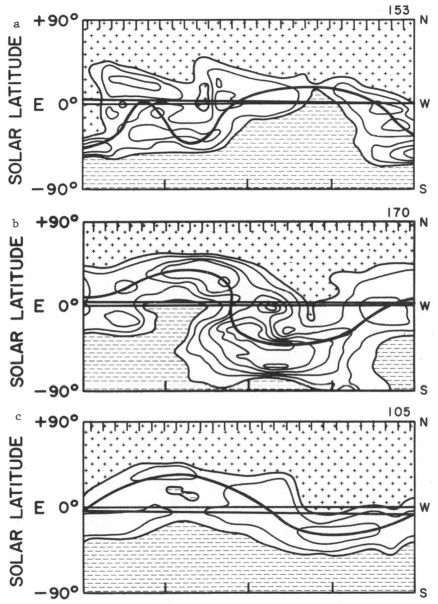

Figure 17: Coronal brightness contour maps (as in Fig. 7) for Carrington rotations 1602, 1616, and 1627. A neutral line location at large distances from the sun is shown in the maps; this line has been drawn through the brightest regions of the corona, constrained to intersect the ecliptic plane in agreement with the observed interplanetary magnetic sectors.

be viewed by a miraculous eyeball (sensitive to coronal, but
not photospheric, radiation, and to distant vector magnetic
fields) located at a Carrington longitude of $\sim 90^{\circ}$, 20° north
of the solar equator is shown in Fig. 18. Evidence was pre-
sented in Section 2 that the neutral line bounding the south
polar cap extension $_1$ was near 25° north latitude. The observation of 750 km sec^{-1} solar wind for portions of rotation 1602
then indicates that θ, the angle from a neutral line occupied
by slower solar wind, is of this same size; $\theta \sim 25^{\circ}$.

The coronal structure for Carrington rotation 1616 was
also discussed in Section 2. The tilted dipole-like character of the large-scale structure for this rotation is now
explicitly illustrated by the neutral line on Fig. 17b; the
brightness map again indicates the presence of other components
in the coronal field, but perhaps to a lesser degree than in
rotation 1602. The brightness map for rotation 1627 (Fig. 17c)
shows a sinusoidal band of bright corona that is remarkably
close to that expected for the corona in an ideal dipole field
(Fig. 16) tilted by about 30°. The three-dimensional density
structure for both rotations 1616 and 1127 should look much
like that in Fig. 15. The tilt angle α implied by most of the
brightness maps from 1974 and 1975 is similar, near 30°. The
presence of two 700–750 km sec^{-1} solar wind streams in nearly
every solar rotation from these years implies a value of the
angle θ slightly less than this tilt angle. Choice of $\theta \sim 25^{\circ}$
is again consistent with the observations.

c. Some Comments on the Phenomenological Descriptions

There should be no grand illusions regarding the profun-
dity of the phenomenological description of coronal and interplanetary structure developed above; it is little more than a
reiteration of the consistency of the data set described in
Section II with unity added to the description by some simple
qualitative concepts regarding the nature of the coronal and
interplanetary magnetic geometry. The magnetic geometries
inferred from observed coronal brightness maps show many features suggested by earlier conceptual models (see below). The
major new feature added in the present discussion is the large
and regular variation of solar wind speed with distance from
magnetic neutral surfaces (or lines) with the uniform, high-
speed, polar cap flow in a major part of interplanetary space.
If this discussion does represent an advance upon earlier
models, the advance is the direct result of the more comprehensive data base used in its formulation, not of any great
advance in physical sophistication. Nonetheless, it is grati-
fying that the observations can be so well coordinated within
so simple an interpretive framework.

N   1602

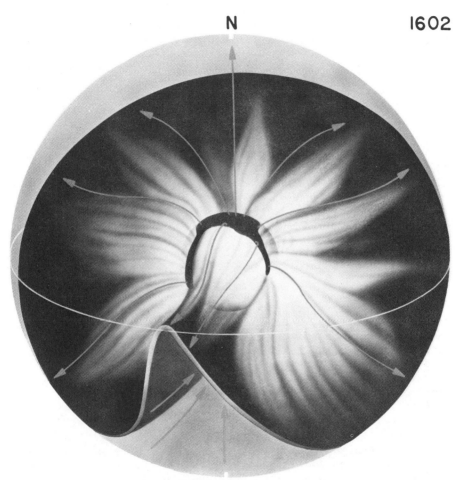

Fig. 18.   An artist's drawing of a hypothetical three-dimensional magnetic and density structure for Carrington rotation 1602, as viewed from Carrington
longitude near 90°, heliocentric latitude 20° N.
The magnetic neutral surface at large heliocentric
distances has been given the shape consistent with
the neutral line on Fig. 17.

The most striking consistency found by applying this model to the 1973-1975 epoch is the resemblance of the inferred coronal structure to that expected in the presence of a dipole-like solar magnetic field (e.g., the three cases in part b above), with this resemblance becoming strongest in early 1975. The possible presence of a dipole component in the complex solar magnetic field should provoke no great surprise. The two polar regions of the sun have been observed to exhibit magnetic fields of opposite polarity through most of the past three solar cycles (with the polarity changing near sunspot maximum); this implied existence of a solar dipole aligned with the axis of solar rotation is part of the widely accepted Babcock-Leighton model of solar magnetic field evolution. The polar structure of the white-light corona observed at eclipse often suggests an open, diverging magnetic structure expected for a dipole-like field. Analysis of measured photospheric magnetic fields throughout much of the past solar cycle (66) and during the Skylab epoch (67) reveal the common presence of a dipole moment in the solar equatorial plane. Such an equatorial dipole may also be implied by the two-sector interplanetary magnetic structure prevalent during this entire epoch (i.e., Fig. 4). The addition of these axial and equatorial dipoles should then lead to a tilted dipole as part of the solar magnetic field.

What may be surprising, however, is the apparent dominant role of this dipole in determining the global structure of the distant corona and the solar wind. The prevailing view of interplanetary magnetic structure during the past decade has laid heavy emphasis on the sector pattern, with its origin in corresponding "solar sectors" with a characteristic north-south orientation in the solar photosphere (e.g., 68-70) epitomized in Fig. 19a. Refinements (e.g., 60,71) of this concept have included combination with the polar or axial dipole as in Fig. 19b, but retained the view of a basically north-south oriented magnetic structure extending over a broad range of latitude. The description set forth above relates the same interplanetary magnetic structure and the associated plasma flow pattern to the observed three-dimensional structure of the corona in a manner that seems to be more closely related to the concepts set forth by Rosenberg and Coleman (52), Rosenberg (53), Schulz (72), Sawyer (54), and Alfvén (73), placing emphasis on a coronal and interplanetary neutral sheet encircling the sun near its equator and thus displaying a basic east-west orientation. Some of this difference in viewpoint could be attributed to the growth of the broad polar extensions during the Skylab epoch; the lane-like coronal hole (hole #1 in catalog (41)) near 0° longitude in Carrington rotation 1602 (see Figs. 8, 17, and 18) may suggest a sector-like structure, but such holes were not common during the

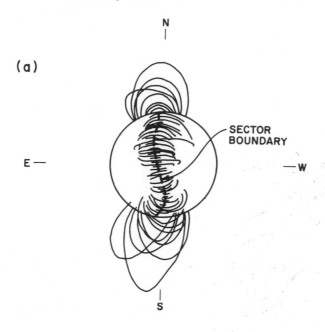

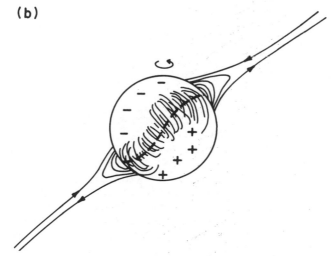

Figure 19: Several earlier views of the
coronal magnetic structure associated with a
magnetic sector boundary (Svalgaard et al.,
1974). Part (a) of the figure shows a north-
south arcade of closed magnetic field loops
extending over a broad range of solar latitude
to form a "solar sector" boundary. Part (b)
shows a combination of this view with a polar
magnetic cap of open field lines.

Skylab epoch. Stated in more conventional terms, the simple
two-sector pattern of the 1973-1975 epoch represents a change
from the four-sector pattern observed in the ecliptic plane
for the earlier portion of solar cycle 20 (and held to be common during most solar cycles by Svalgaard and Wilcox (58)).
Some difference could also stem for emphasis on the neutral
line geometry at the photosphere in the sectarian view of
Wilcox and colleagues, and on the outer corona in the present
description. The influence of smaller-scale features (or
"higher harmonics") in the solar magnetic field should diminish
more rapidly with altitude than that of large-scale features
such as a dipole, leading to an increasing dominance of the
latter at greater heights (e.g., 74,75). Nonetheless, this
author regards the views concerning coronal magnetic geometry
discussed here, in Svalgaard (5), and in Svalgaard and Wilcox
(76) as a re-orientation toward the view advocated in the
early and often neglected work of Rosenberg and Coleman (52),
and regards the evidence for a near-equatorial interplanetary
neutral surface as highly convincing. Additional supporting
evidence has recently been reported by Smith et al. (77) from
Pioneer spacecraft observations made at a solar latitude of
$\sim 16^\circ$. The true geometry of the interplanetary magnetic field
is, of course, not uniquely determined by the observations and
phenomenological arguments invoked here; it will probably
remain poorly determined until extensive in situ observations
are made out of the ecliptic plane.

The three-dimensional pattern of solar wind flow that we
have associated with this magnetic structure represents a
major revision of earlier views. Much attention has been
focused in the past on the development of quantitative models
of the coronal expansion capable of explaining a "quiet" or
slowly varying state of the solar wind. This state has often
been identified (see Chapter III of (27)) with conditions prevailing when the solar wind speed is low. The extremely fast
solar wind observed in large solar wind streams in 1973 and
1974 is found to satisfy many criteria for a "quiet" state
(78). Our inferrence that this type of wind exists outside of
a "belt" within 25° of a single interplanetary neutral sheet
implies that over half of the interplanetary region is pervaded
by this very fast flow in 1973 through 1975. Any complete
model of the expanding corona must be capable of explaining
this newly observed extremely energetic state of the solar
wind.

The generality of our conclusions regarding the spatial
variation in solar wind speed remains unproven. Attempts to
infer such variations in geomagnetic activity over the $\pm 7^\circ$
annual solar latitude excursion of the earth have long been
controversial (see 23,79). The existing level of semi-annual

geomagnetic fluctuations can also be attributed to the changing
tilt of the terrestrial rotation (and hence magnetic) axis;
in (5) Svalgaard argues against any heliographic latitude gradient in the solar wind greater than 1 or 2 km sec^{-1} degree^{-1}
on the basis of this interpretation. Solar wind speeds inferred from comet-tail observations performed over 75 years show
no significant speed variation within $\sim 50°$ of the solar equator (80). Semi-annual fluctuations in directly-observed solar
wind speeds have been found (48) but only for several years in
the past solar activity cycle (81), again indicating no large
latitude gradients to be generally present. All of this evidence might be regarded as indicating that the steep speed
gradients advocated here (see also 82) and in earlier studies
(e.g., 83) are not common. Any such conclusion is unjustified;
even the strong gradients implied in our phenomenological model for 1974-1975 do not lead to large variations in the solar
wind speed, averaged over longitude, at latitudes within $\pm 40°$
of the solar equator. It is, in contrast, intriguing to find
that the average speed gradient normal to the neutral surface
of the present model, 17 km sec^{-1} degree^{-1} agrees well with
the value of ~ 15 km sec^{-1} deduced by Rhodes and Smith for a
1967 epoch.

d. Variations in Coronal and Interplanetary Structure During the Solar Activity Cycle

The observed characteristics of coronal holes and the
associated interplanetary sectors and streams during the 1972-
1975 interval suggest a systematic change in coronal structure.
In 1972 and 1973 an evolving pattern of holes, with more than
two such features generally present, produced an evolving
interplanetary two-sector pattern with more than a single high
speed stream occurring within the sector of positive magnetic
polarity. In 1974 and 1975 a nearly stationary pattern of two
large, polar-connected holes produced a highly recurrent and
tightly synchronized two-sector, two-stream structure. Correlated with the latter were a pair of recurrent geomagnetic
disturbances that persisted for the two years and suggest the
growth of M-regions and recurrent geomagnetic activity common
to the declining phase of many earlier solar cycles.

Any extension of this simple statement concerning the
evolution of the corona and solar wind over any longer period
of time must be speculative. No coronal data base of similar
quality or completeness exists for the earlier phases of solar
cycle 20; see (84) for a discussion of X-ray and XUV observations from earlier phases of solar cycle 20 and their implications with respect to coronal evolution. Further, there is
evidence that the M-region phenomenon for this cycle was of
exceptional intensity (see 5 and 85). Nonetheless, the

availability of direct solar wind observations for much of
this cycle (on a regular basis since ∿ mid-1965) does provide
a partial data set (e.g., 22,86). A general understanding of
the evolution of the large-scale solar magnetic field during
a solar cycle permits some qualified deduction of the nature
of coronal fields and hence structure that can explain, in a
very general sense, the observed changes in solar wind properties. Despite its speculative nature, we will briefly describe this extension of the phenomenological models of coronal
structure beyond the Skylab epoch.

A logical starting point in the evolution of the solar
magnetic field is the reversal of polar magnetic fields at a
time near the maximum in the activity cycle. Following the
reversal in solar cycle 20, an outward-pointing field in the
north polar region and an inward-pointing field in the south
polar region is expected to grow in intensity throughout the
remainder of the cycle; see Sheeley (87) for estimates of
these intensities from counts of polar faculae. At the same
time the magnetic fields in low and mid-latitudes should become more organized, changing more slowly with time and
exhibiting a dominant magnetic polarity over larger spatial
scales. These traits of the solar magnetic field should lead
to
(1) the existence of an axial dipole structure with polar-
     cap holes,
(2) the existence of low-latitude holes of increasing size
     and lifetime.
The combination of these two classes of open magnetic structures might then produce the general trend in coronal structure illustrated in Fig. 20. Some time after solar maximum,
small polar holes and small, isolated, and short-lived low-
latitude holes would be expected (part a of Fig. 20). The
associated interplanetary structures observed in the ecliptic
plane would then tend to be short-lived (or not to recur for
many solar rotations), and the high-speed streams to be of
small amplitude. At a later time both classes of holes should
become more stable and larger in size, thus tending to "connect" to the polar cap hole of the same magnetic polarity,
forming the "lane" type of coronal holes observed in the years
before Skylab (e.g., 3,34) and during the first half of the
Skylab mission (part b of Fig. 20). Near the end of the solar
cycle, this same trend should lead to the appearance of broad
"polar extensions" and very long-lived, large-scale distor-
tions of the basic axial dipole configuration (Fig. 20c). In
all of these evolutionary stages, the corona should have had
the general appearance often noted at eclipses near sunspot
minima--faint polar caps and bright equatorial extensions.
In solar cycle 20, the low latitude fields developed a strong
equatorial dipole component (see 67) leading to the simple

tilted dipole configuration of the outer corona in 1974 and
1975. The associated solar wind structure in this stage of
evolution should then be simple, strongly recurrent, and
characterized by the appearance in the ecliptic plane of very
rapid flows related to the polar extensions. These changes in
solar wind properties correspond to those observed in the declining portion of solar cycle 20 (22,40,44,47).

The beginning of a new solar cycle should lead to a rapid
breakdown of the ordered magnetic structure at mid and low
latitudes and the slow erosion of the polar cap fields. One
should then expect a persistence of the axial dipole and its
associated coronal structure well into the new cycle, but a
rapid disappearance of polar cap extensions and large coronal
holes (Fig. 20d). The general appearance of the corona should
still be that seen at eclipses near sunspot minima but with
symmetry about the solar rotation axis. The prominent recurrent sectors, streams and geomagnetic activity sequences
should end abruptly (see 5). The axial dipole structure could
well lead to the existence of an interplanetary neutral sheet
very close to the solar equator. The solar wind in the ecliptic plane would then show the generally low speeds common in
proximity to the neutral sheet and the magnetic polarity would
display a strong dependence on solar latitude. Recurrence
with the 27-day solar rotation period should become rare. If
the neutral sheet were in fact within 7° of the solar equator
at the orbit of earth, the annual excursion of the earth in
solar latitude could lead to the "monopolar" sector structure
noted by Wilcox (88) after recent solar minima. In any case,
six month periodicities in solar wind structure (52,89) and
geomagnetism should be expected. The growth of activity in
the new cycle should disrupt this pattern as the polar cap
magnetic fields decline in intensity and stronger, short-lived
fields disrupt the corona at low latitudes. Transient disrup-
tions of the solar wind, the largest of which are related to
flares (chapter VI of 27) should play a role of increasing
importance as the old order established in the previous cycle
(and observed for solar cycle 20 during the Skylab mission)
fades away. The state of the corona during the interval when
the polar fields reverse sign, and the magnetic polar caps
might temporarily disappear, should be quite interesting. It
falls, however, even beyond the weak chain of inference used
here and must await future studies. So too must the applicability of this model to other solar cycles.

e.   The Evolution of Solar Magnetic Fields in 1972 and 1973

The pattern of coronal hole evolution between mid-1972
and the end of 1973 has been described in Section 2. Briefly,
holes of a given magnetic polarity tended to appear, mature,

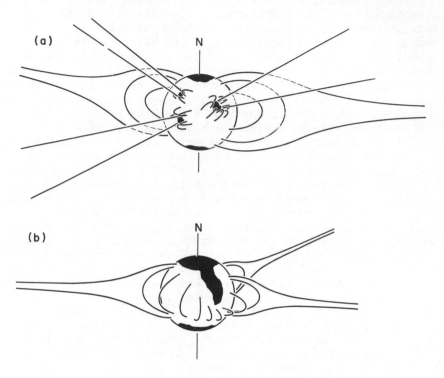

Figure 20   A model for the temporal evolution
of the coronal structure during the solar
activity cycle.   Parts b and c show the structure
inferred from observations early in the Skylab
mission (e.g., rotation 1602) and the stable
tilted dipole of 1974 and early 1975 (e.g.,

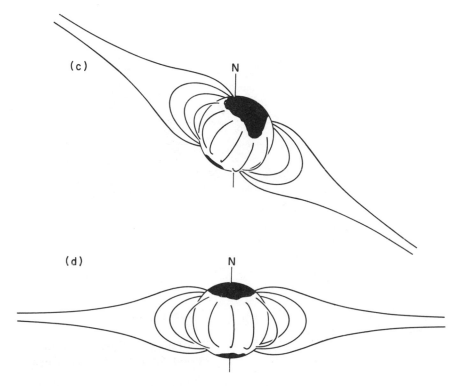

rotations 1616 and 1627).   An earlier phase,
with smaller polar caps and several small, isolated holes is suggested in part a, and an
axial dipole, part d is suggested for the post-
sunspot minimum state of the corona.

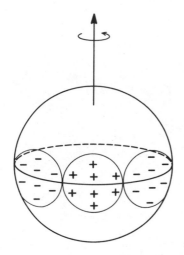 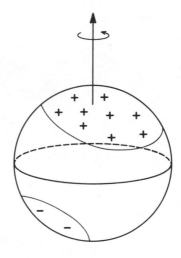

"Three-Wave" Polarity
Structure Near Equator

Tilted Dipole

Rotates With 27 Day Period

Rotates With Global Average
Period, Greater Than 27 Days

Figure 21: A schematic two-component decomposition of the magnetic field at the base of the corona in early 1973.

and disappear in a systematic pattern of 90° to 120° eastward
jumps; individual holes and the associated streams appear
nearly stationary in a frame of reference rotating with a period near 27 days (e.g., Carrington or Bartels solar rotations)
but the magnetic polarity pattern drifts eastward or displays
a longer recurrrence period. If the generally assumed relationship of coronal holes and the large-scale solar magnetic
field holds for these features, the magnetic field must display a similar structure in solar longitude.

A simple explanation of the coronal holes with positive
magnetic polarity can be given by assuming the solar magnetic
field to consist of two "components", as shown schematically
in Fig. 21. One of these represents a pattern of outward and
inward pointing fields near the solar equator; in the figure
a "three-wave" structure at the equator is used for simplicity. The other represents a tilted dipole as inferred from
the observed coronal structure during the Skylab epoch, or
a larger-scale magnetic structure. The smaller-scale magnetic
features would be related to electrical currents near the
solar equator and near the surface of the sun, and should thus
rotate with these layers. In contrast, the larger-scale or
dipole pattern would be related to electric currents distributed widely over solar latitude and possibly extending to
some depth beneath the surface. It must then rotate at a
rate characteristic of a large portion of the atmosphere, or
at the average rotation rate of its widely distributed sources.
In the differentially rotating sun, this average rate must be
slower than the equatorial rotation rate. Hence the dipole
field pattern of Fig. 21 should drift eastward relative to
the small-scale structures.

The sum of these two magnetic field "components" will
then depend on their shifting relative phase. The region of
open magnetic field lines associated with a smaller-scale
area of given magnetic polarity should grow, connect to the
appropriate polar cap, and diminish as the equatorial dipole
component of the same polarity drifts into and out of its
longitude. Each such individual open region or hole will
remain essentially fixed in the frame of reference rotating
with the near-equatorial rate. This evolving solar magnetic
and coronal structure would lead to high-speed solar wind
streams that show a similar pattern of growth and decay, and
the same rotation rate, as the low-latitude holes. The interplanetary magnetic polarity pattern near the solar equator
would, however, be dominated by the dipole pattern, rotating
with the longer global average rotation rate.

This simple explanation (see also 4) has been contrived
to produce the observed pattern of coronal and interplanetary

evolution in mid-1972 through 1973 and thus its success may not be surprising. Stix (90) has derived special solutions of the dynamo equations that represent wavelike longitudinal magnetic structures whose rotation lags behind the equatorial rotation for long wavelengths, illustrating the physical plausibility of the difference in rotation rates invoked above. Further, the spherical harmonic analysis of measured photospheric magnetic fields by Stix (67) does indicate that a dipole and a "three-wave" equatorial structure are the dominant long-wavelength components present in 1972-1973. The description given above, though greatly oversimplified, may thus bear some resemblance to reality, and the solar magnetic field may indeed display rather distinct modes or waves standing or drifting coherently in its differentially rotating atmosphere.

The abrupt change in the evolution of coronal structure near the end of the Skylab mission is also interesting in this context; the increasing resemblance of the coronal brightness maps to the pattern expected for a tilted dipole (part b above) suggests a growth of the dipole component in the solar magnetic field. Indeed, Stix (67) does find a large increase in the equatorial dipole component of the measured photospheric fields in 1974. If, however, we continue to regard the dipole rotation rate as a global average, we would be forced to conclude that the associated average period decreased by more than a day in late 1973 and early 1974. Howard (91) has, in fact, recently reported observed solar rotation rates that change over the past decade and, in particular, imply a decrease in rotation period by slightly more than a day over a wide range of solar latitudes between 1973 and 1974.

4. Physical Implications and Problems

The phenomenological descriptions set forth in the previous section are but initial steps toward an understanding of the observed coronal and interplanetary structures. Our ultimate goal is physical understanding; for example, not simply the knowledge that observed high-speed solar wind emanates from coronal holes, but the understanding of the cause of high-speed flow from such regions. Unfortunately, the leap from phenomenological models to quantitative models based on physical laws is a difficult one. Despite the relative simplicity of coronal and solar wind structures during the Skylab epoch, quantitative modeling of these structures remains a formidable task. This section will be an attempt to focus attention on some basic implications and particular problems posed for modeling efforts by the observations from the Skylab epoch.

a.   The Magnetic Geometry Associated with Coronal Holes

Perhaps the strongest implication of this body of observations is one readily apparent in Skylab data -- the high
degree of coupling between coronal and interplanetary structure and the solar magnetic field. The simplest view of the
result of this coupling is probably that illustrated in Fig.
2 where coronal holes, observed as dark regions in emission
from the low corona or in scattered light from the outer corona, are identified with the long-lived open regions in the
large-scale coronal magnetic geometry. This identity implies
that the low speed portions of solar wind streams also emanate from coronal holes, representing the coronal expansion
along the magnetic flux tubes near the edges of the open regions. This view is attractive to many because of its relative simplicity and because of its correspondence to two features of existing physical models; namely,

(1)   the difference between the static equilibrium on closed
    magnetic field lines and the steady outflow on open mag-
    netic flux tubes leads naturally to sharply bounded high
    and low density regions associated with these magnetic
    structures (e.g., 65):

(2)   the inferred modulation of expansion speed across an open
    magnetic region -- low speed near the edges and higher
    speed from the center -- is predicted quite generally for
    a conduction-dominated coronal expansion (e.g., 92).

It has thus been the starting point for most efforts to model
the coronal holes and the associated flow of solar wind.

Unfortunately, the relationship between coronal holes and
magnetic geometry is only suggested by observation, and the
high degree of correlation between holes and the interplanetary sector-stream structure does not prove that all of the
plasma and magnetic field in the latter came from the observed
solar features. Thus the identity of coronal holes and the
large-scale open magnetic regions of the corona must be regarded as a hypothesis, subject to observational confirmation.
For the sake of completeness several other possible hole-
magnetic geometry-solar wind relationships should be mentioned
here.

(1)   The solar wind emerges from long-lived open regions in
    the large-scale magnetic field as before. However, the
    differences in the density structure within a given open
    region are large, with only a portion of that region
    (most likely the central portion) so tenuous as to be
    identified as a coronal hole. This hypothesis eliminates
    the identification of the sharp boundaries of observed
    holes with the sharp demarcation between open and closed
    magnetic regions (on the large scale).

(2)   The solar wind emerges both from large-scale open regions

and from smaller regions of open magnetic field lines
scattered within the neighboring, largely closed portions
of the corona. The small source regions would thus fall
outside of the observed boundaries of coronal holes, and
must give rise to slow-speed solar wind. The magnetic
polarity of the small regions must exhibit the large-
scale organization implied by the interplanetary sector
structure.

(3) The boundaries between large-scale open and closed magnetic regions of the corona are the sites of numerous,
small transient emissions of plasma and magnetic field.
Solar wind would thus emanate from the large-scale open
regions much as before and from a "fuzzy" boundary region, most of which would not be identified as belonging
to a coronal hole at any given time.

(4) The coronal magnetic field is not strictly frozen into
the coronal flow. Cross-field diffusion would most logi-
cally occur from dense, closed regions to add to the
plasma flow from the open regions.

All of these possible configurations would lead to the same
correlation between observed coronal holes and the interplanetary stream-sector structure. All could be grafted onto the
phenomenological descriptions of Section 3 with no fundamental changes. They would, however, add serious complexity and
difficulty to any quantitative models of these phenomena.

b. Magnetic Fluxes in the Corona and Solar Wind

Any proposed mapping of a steady coronal source region
into an associated interplanetary structure is subject to a
quantitative test that is utterly simple in principle. Suppose, as in Fig. 22, that some region of the corona, such as
a coronal hole, is thought to map into a solar wind feature
observed at 1 AU. Then the rate at which a conserved physical
quantity flows through the source region must equal that at
which the same quantity flows through a sun-centered sphere
within the associated solar wind feature (under the steady
flow assumption). Thus, for example, the magnetic flux at
the base of a coronal hole

$$\Phi_b = \int^{hole} B_r \, dA$$

with the integral of the radial field component B_r taken over
the entire area of the hole, must equal the interplanetary
flux

$$\Phi_i = \int^{s.w.feature} B_r \, dA$$

with the integral now taken over the area of the solar wind
feature on the sun-centered sphere. The integrals can, of

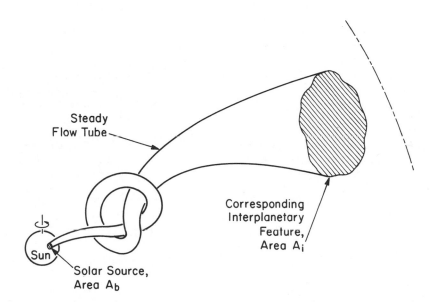

Figure 22: The mapping of fluxes by a steady flow from a coronal source region to its interplanetary counterpart. The particular flow tube geometry illustrated here has been carefully selected to emphasize the generality of the flux conservation argument.

course, be extended to the entire sphere in interplanetary
space and the entire area of all postulated solar wind sources
(such as holes) at the base of the corona.

Application of this test turns out to be far from simple
in practice. The line-of-sight magnetic field component in
the solar photosphere is routinely measured (e.g., 4) and
provides a suitable integrand B_r at the base of the corona.
The fields beneath low or middle latitude coronal holes at
the time of Skylab were generally weak and thus particularly
subject to observational errors (for example, the common difficulty with the zero level of magnetographs). The polar cap
magnetic fields are also weak, and determination of the radial
component from observations directly pertinent to the line-
of-sight component adds an additional uncertainty. Nonetheless, the observed B_r can be integrated over coronal holes to
yield a value of Φ_b in which the major errors will stem from
the observations of B_r. The vector magnetic field is routinely measured by in situ probes in interplanetary space (e.g.
93). Unfortunately, these observations from a single spacecraft yield the variation in $\vec{B}$ as a function of time or longitude on a line crossing the entire area occupied by a solar
wind feature. As virtually all spacecraft orbits are in or
very near the ecliptic plane, integration over this area requires some assumption regarding the variation of B_r away
from that line (or basically with solar latitude). Fortunately, the intensity of the magnetic field away from the compressed regions at the leading edges of high-speed streams is
one of the least variable properties of the solar wind, and
the orientation of $\vec{B}$ on time-scales longer than a few hours
is well organized into the spiral pattern. It is thus reasonable to assume that B_r does not vary rapidly with solar latitude. This assumption gains further credibility from a quantitative model of plasma flow from a polar hole (62), indicating that meridional flows in interplanetary space tend to
lead to a uniform $|\vec{B}|$. Then

$$\Phi_i \sim <B_r>_i \, A_i$$

where $<B_r>_i$ is the average magnetic field observed across the
solar wind feature and A_i is its area on the sun-centered
sphere. The area A_i must then be inferred in some manner,
and the major errors in our estimate of Φ_i will stem from extending the accurate measurements of $\vec{B}$ over this poorly known
area. The equality of the coronal and interplanetary fluxes
then implies that

$$<B_r>_i \, A_i = <B_r>_b \, A_b$$

where Φ_b has been written as the product of the average radial magnetic field $<B_r>_b$ over the area A_b at the base of the coronal hole. Thus the average magnetic fields at the base of the corona and in interplanetary space must be related as

$$<B_r>_b = <B_r>_i \frac{A_i}{A_b}$$

Table 1 shows the values of average coronal magnetic fields inferred in this manner for several large solar wind streams observed during the Skylab mission (as drawn from the tabulation of coronal holes and streams in Table 1 of (1)). In each of these cases $<B_r>_i$ was determined from in situ observations and the area ratio A_i/A_b estimated from the relative widths in longitude of the associated magnetic sector observed at 1 AU and the coronal hole with which the specific stream was correlated, or from an expansion factor for polar holes deduced by Munro and Jackson (94) from Skylab observations in July 1973. Values of $<B_r>_b$ ranging from 6 to 12 gauss are obtained. A similar value results from assuming an average interplanetary $<B_r>_i$ of 4×10^{-5} gauss over the entire sphere and assuming that coronal holes occupy 20% of the base of the corona (1 and 3):

$$<B_r>_b = 8 \text{ gauss.}$$

Burlaga et al. (95) have deduced similar values from a more extensive study of streams and holes based on data from 1974. Thus the hypothesis that individual coronal holes are identical to the set of open magnetic field lines giving rise to individual solar wind streams, or as a collection giving rise to all of the solar wind, leads to the conclusion that the average radial magnetic field strength at the base of coronal holes must be close to 10 gauss.

These inferred values of average magnetic field strengths are significantly higher than those usually given for the "background" field in the solar photosphere (96) or the typical fields in the photospheric regions of dominant magnetic polarity traditionally correlated with the interplanetary sector pattern (4). A further exploration of this problem in the context of potential models of the coronal magnetic field is described in (4). In essence, this test of the "identity hypothesis" seems to lead to a disparity by factor of 3 or more in the magnetic fluxes inferred in the solar wind and in coronal holes. Possible sources of this disparity are:
(a) systematic errors in measurements of the weak photospheric magnetic fields underlying coronal holes,
(b) lack of knowledge of strong magnetic fields in the polar regions of the sun,
(c) overestimation of the interplanetary flux due to its concentration near the solar equator.

However, if the disparity is real, the identification of coronal holes with open magnetic regions has led to a serious
difficulty. All of the alternate coronal structures mentioned
in part a of this section except for possibility (4) would
help to alleviate the disparity by permitting the interplanetary magnetic field to be rooted in a larger area of the sun
than that occupied by observed coronal holes.

c. Particle and Energy Fluxes in the Corona and Solar Wind

The same principle used above to relate the coronal and
interplanetary magnetic flux implies the equality of the particle flux

$$I = \int nu_r \, dA \ ,$$

(where n is the proton or electron density and u_r is the radial expansion velocity component), and the energy flux

$$W = \int w \, dA \ ,$$

(where w is the energy flux per unit area in the flow), integrated on sun-centered spherical surfaces, for any steady flow
tube or related coronal and interplanetary features. The
scarcity of observations of the outward expansion velocity in
the visible corona precludes using these integrals as tests of
any postulated coronal-interplanetary mapping of features.
However, estimates of particle or energy flux based on interplanetary observations imply the coronal values required to
maintain the steady solar wind flow (in a particular feature
or for the entire solar wind). These inferred fluxes can
serve as a guide to understanding the mass and energy balance
of the corona and as important constraints on quantitative
models of the expanding corona.

Table 2 summarizes particle and energy flux estimates
for the same set of high-speed solar wind streams used in the
discussion of magnetic flux. The observed average particle
flux densities range from 3 to 4.5 x 10^8 protons cm^{-2} sec^{-1}
while average energy flux densities (including the work done
against solar gravity) range from 1.8 to 2.7 ergs cm^{-2} sec^{-1}
at 1 AU for these features. Feldman et al. (97) have obtained
similar values from a more extensive study. Mapping of these
observed values back to coronal holes by the same correspondence of areas at the base of the corona and at 1 AU implies
average fluxes at one solar radius of

$$\langle nu_r \rangle_b = 6 \text{ to } 11 \times 10^{13} \text{ protons cm}^{-2} \text{ sec}^{-1} \ ,$$

$$\langle w \rangle_b = 3 \text{ to } 9 \ \times 10^5 \ \text{ergs cm}^{-2} \text{ sec}^{-1}$$

TABLE TWO: PARTICLE AND ENERGY FLOWS IN STREAMS AND HOLES

EXAMPLE 1: (as in Table 1)

Average fluxes at 1 AU:

Proton flux $= 3.8 \times 10^8 \, \text{cm}^{-2} \, \text{sec}^{-1}$ in "flat region" near peak of stream

<Proton flux> $= 4 \times 10^8$ over stream

Energy flux $= 2.1$ ergs $\text{cm}^{-2} \, \text{sec}^{-1}$ in "flat region"

<Energy flux> $= 2.1$ ergs $\text{cm}^{-2} \, \text{sec}^{-1}$ over stream.

Transforming to $R_\odot$ using same area ratio as in Table 1 gives the following average values in the coronal hole:

<Proton flux> $= 9.6 \times 10^{13} \, \text{cm}^{-2} \, \text{sec}^{-1}$

<Energy flux> $= 5 \times 10^5$ ergs $\text{cm}^{-2} \, \text{sec}^{-1}$.

EXAMPLE 2:

Average fluxes at 1 AU:

Carrington Rotation	$\langle$Proton flux$\rangle$	$\langle$Energy flux$\rangle$
1598	3.5×10^8 cm^{-2} sec^{-1}	2.6 ergs cm^{-2} sec^{-1}
1599	3.3	2.6
1600	3.8	2.9
1601	3.5	2.7
Averages	3.5×10^8	2.7

Transforming to $R_\odot$ using the area ratios of Table 1 gives the coronal values:

$\langle$Proton flux$\rangle = 1.1 \times 10^{14}$ cm^{-2} sec^{-1}

$\langle$Energy flux$\rangle = 8.7 \times 10^5$ ergs cm^{-2} sec^{-1}

EXAMPLE 3:

Average fluxes at 1 AU:

$\langle$Proton flux$\rangle$ = 4.4 x 10^8 cm^{-2}sec^{-1}

$\langle$Energy flux$\rangle$ = 2.2 ergs cm^{-2}sec^{-1}.

Transformation to R_o using the area ratio from Table 1 gives the coronal values:

$\langle$Proton flux$\rangle$ = 6.3 x 10^{13} cm^{-2} sec^{-1}

$\langle$Energy flux$\rangle$ = 3.2 x 10^5 ergs cm^{-2} sec^{-1}.

EXAMPLE 4:

Average fluxes at 1 AU:

$\langle$Proton flux$\rangle$ = 4.0 x 10^8 cm^{-2} sec^{-1}

$\langle$Energy flux$\rangle$ = 1.8 ergs cm^{-2}sec^{-1}.

Transformation to R_o using the area ratio from Table 1 gives the coronal values:

$\langle$Proton flux$\rangle$ = 9.6 x 10^{13} cm^{-2}sec^{-1}

$\langle$Energy flux$\rangle$ = 4.4 x 10^5 ergs cm^{-2}sec^{-1}.

in the coronal holes giving rise to these streams. The particle fluxes in the monster stream and the two large streams of
1974-1975 are only 15 to 50% higher than the average values
inferred from studies earlier in this solar cycle. Thus the
total particle flux in the solar wind is raised by less than
25% if half of interplanetary space is occupied by such streams
(Section 3). Using $<nu_r> = 3 \times 10^8$ protons cm^{-2} sec^{-1} for
"normal" solar wind ($\underline{89}$) and $<nu_r> = 4.5 \times 10^8$ protons cm^{-2}
sec^{-1} in the large streams gives

$$I \approx 1.1 \times 10^{36} \text{ protons sec}^{-1}$$

(or a mass loss rate of about 2×10^{12} grams sec^{-1}). Feldman
et al. ($\underline{97}$) report somewhat different fluxes in both high and
low-speed wind from a more extensive study of data from the
1971-1974 epoch. These values lead to a lower value of
$I \approx 9 \times 10^{35}$ photons sec^{-1}. However the energy fluxes in
these streams are several times higher than the earlier averages, and their presence significantly changes our estimate
of the total energy carried by the solar wind. Using $<w> \approx 1$
erg cm^{-2} sec^{-1} for the normal wind and $<w> \approx 2$ ergs cm^{-2} sec^{-1}
in the large streams gives

$$W = 4 \times 10^{28} \text{ ergs sec}^{-1}$$

as the energy flow in the solar wind.

The estimated particle flux density of $\sim 10^{14}$ protons
cm^{-2} sec^{-1} in coronal holes implies a much more rapid expansion of the low corona than indicated either by earlier estimates assuming flow from a large part of the corona or by
theoretical models of a spherically symmetric coronal expansion. For example, a coronal density of 5×10^7 electrons or
protons at ~ 0.1 solar radii above the base of the corona (a
reasonable extrapolation from ($\underline{94}$)) implies an expansion speed
of 20 km sec^{-1}; earlier estimates and the theoretical models
generally gave a few km sec^{-1}. Any modification of the assumed
source areas for the streams would, of course, change the
inferred coronal expansion speed.

The estimated energy flux density of $\sim 5 \times 10^5$ ergs cm^{-2}
sec^{-1} is more than an order of magnitude higher than old estimates (e.g., $\underline{27}$) due both to the mapping to a small fraction
of the corona and the larger interplanetary energy flows
observed in the monster type streams. The new value has two
important implications. First, it is a severe requirement
upon any mechanism assumed to transport energy in the low
corona where expansion speeds are low. For example, considerable effort has been expended on coronal expansion models in
which electron heat conduction is assumed to be the dominant

energy transport mechanism in the low corona. However, this
mechanism cannot supply energy at the rate deduced here unless
the temperature in coronal holes is above 2×10^6 °K. Although
we have little knowledge of the coronal temperatures in holes,
they are believed to be lower (and observed to be so at very
low latitudes) than in closed magnetic regions. As the temperature inferred from coronal emission line studies and hence
pertinent to dense or closed regions is about 1.5×10^6 °K, a
2×10^6 °K value in coronal holes may well be unacceptably
high. It can thus be argued that some energy transport mechanism other than electron heat conduction is required to drive
the solar wind. Second, the overall energy balance of the
corona is significantly changed if 5×10^5 ergs cm^{-2} sec^{-1} is
required for the expansion from coronal holes; the solar wind
must then be the dominant energy loss from such regions (see
2).

d. Some Specific Implications with Respect to Quantitative Models

Virtually all contemporary efforts to derive quantitative
models of the solar corona from physical principles have been
based on the equations of magnetohydrodynamics (for a discussion of such models see (2)). Unfortunately these equations
are of such complexity that solutions have been obtained, even
with the aid of modern computers, only for the simplest of
physical assumptions and boundary conditions; the most realistic treatment of an expanding mhd corona has been for an isothermal plasma with a dipole magnetic field and constant density at the base of the corona (29,64), as described in Section
3a. More realistic physical assumptions and magnetic geometries
have been modeled only at the sacrifice of some aspect of
mhd self-consistency. In particular, the computation of the
outward flow along prescribed, open magnetic flux tubes has
been widely used in modeling the coronal expansion from coronal holes; the magnetic field can either be given an assumed,
idealized form in studies of general effects (98-100) or
derived from "realistic" boundary conditions using some
approximation short of the full mhd solution (92,101). In
either case, the resulting plasma flow and magnetic configuration does not satisfy the momentum equation transverse to
the magnetic field lines.

Such models have yielded some valuable insights into
effects of the magnetic geometry on the expanding corona.
Perhaps the most important of these is a simple explanation
for the apparent modulation of solar wind across an open magnetic region -- the fast wind from the centers of coronal
holes, slow wind from near the edges of holes correlation described in Sections 2 and 3. If electric currents are neglec-

ted in the corona, the magnetic field can be derived from a
scalar potential, $B = - \nabla\phi$, with $\nabla^2\phi = 0$ for any distribution
of the radial or line-of-sight magnetic field at the base of
the corona. The ultimate drawing of solar magnetic field
lines into interplanetary space by the solar wind can be simu-
lated by specifying an equipotential spherical surface, at
which all magnetic field lines become radial, at some reasonable distance from the sun (102,103), usually near 2 solar
radii. Computation of such a "potential field" approximation
in the corona, using measured photospheric magnetic fields as
a boundary condition, gives a magnetic geometry in which the
magnetic flux tubes diverge more rapidly with distance from
the sun near the edges of open regions (4) than near the
centers of open regions (see (4) for specific examples).
Integration of the mass, momentum, and energy conservation
equations along open field lines under the assumption that
heat conduction along the field lines is the dominant energy
transport mechanism in the corona then yields a faster coronal expansion in the less rapidly diverging central portions,
a slower expansion in the more rapidly diverging edges of
open magnetic regions (65,92,104,105). This effect, proposed
as the explanation of observed modulations of the solar wind
speed by Billings and Roberts (28), Hundhausen (27), and
Pneuman (65) appears to stem from the greater "spreading out"
of a field line-guided energy flux (such as heat conduction)
on more rapidly diverging lines. Further examples of the
anticorrelation of solar wind speed with flux tube divergence
are given by Levine (4).

There are, however, some very important failings in these
models. The interplanetary magnetic flux implied by the potential field computations is generally well below that inferred from in situ observations. The areas of open regions
given by the potential approximation appear to be smaller than
those implied by coronal hole observations unless the equipotential surface is placed very low in the corona (4).
These problems, obviously related to the broader difficulties
incurred in comparing the coronal and interplanetary magnetic
fluxes above, cast some doubt on the quantitative correspondence of potential fields to the actual coronal magnetic
field. The solar wind density and speed predicted by the flow
solution along these field lines are generally much lower than
observed. In other words, the simple physical assumption of
energy transport by heat conduction is inadequate to lift
sufficient mass and carry sufficient energy away from the
sun to explain the observed solar wind.

This latter problem can be related to a general difficulty encountered in solar wind models for the past decade.
The complexity and nonlinearity of the equations of ordinary

fluid mechanics as well as magnetohydrodynamics has always
dictated a difficult choice in the extension of the theory of
the coronal expansion. The additon of either geometric or
physical complexity to simple models pushes the theory to the
ragged edge of mathematical tractability. Thus the theory has
usually been extended by adding physical processes in the
simplest of geometries, that of a steady, spherically symmetric, and often radial coronal expansion. Much of the emphasis in this past work concerned a disparity between the solar
wind speeds of $\sim$ 250 to 300 km sec^{-1} predicted by most models
and the observed speeds of 300 to 400 km sec^{-1} thought then to
be typical of "quiet" solar wind (see (27), Chapter III).
Much less interest was shown in more complex geometries. However, the simplified models described above indicate that a
magnetic geometry not unlike that inferred from coronal observations can have 100% effects on the coronal expansion. These
effects include a general reduction in the expansion speed, a
regrettable circumstance in view of the observation during the
Skylab epoch of "quiet" solar wind with unprecedented 750 km
sec^{-1} flow speeds (Section 3).

We are thus led to a rather sobering view of the present
state of quantitative models of the coronal expansion in holes.
All existing models correspond to the simplest of possible
relationships (Fig.2) between the coronal magnetic field and
the expanding plasma--steady, field-aligned flow in a simple
large-scale magnetic geometry. Any modification of this relationship required to solve the "flux problem" of part B above
would be a substantial complication (in boundary conditions at
the very least) of modeling efforts. Comparison of existing
models with observations made during the Skylab epoch, and in
particular with the fast, energetic flows in the large-
amplitude solar wind streams of 1973 and 1974, reinforces
earlier arguments for significant energy or momentum addition
to the solar wind by mechanisms other than heat conduction in
the corona (see 106-109). Incorporation of such effects into
models not only increases their complexity, but divorces them
from presently observable boundary conditions. Further, these
physical modifications cannot long be studied in isolation
from the geometric difficulties introduced by flow in a diverging magnetic field. The import of the latter has been well
demonstrated by Kopp and Holzer (98) who have integrated the
mass and momentum conservation equations for a polytropic
flow on a radial flow line in a rapidly diverging flow tube.
The diverging geometry lowers the region of greatest acceleration of the expanding plasma (for solutions with a constant
flow speed at large heliocentric distances), and thus increases
the effects of any physical processes occurring in the lower
corona. Sufficiently rapid divergence can even lead to a

discontinuous change in the basic topology of the solutions
to the fluid equations (110).

Although sobering, this view should not be regarded as
hopelessly discouraging. The advances in phenomenology described above could come quickly given the superb data set
available from Skylab. It should not be surprising that similar advances in physical understanding and quantitative modeling follow at a slower pace. The added geometric complexities
that now appear to be essential features of the coronal expansion require re-examination of the multitude of physical processes suggested as cures for the ills of earlier models.
While this re-examination is not trivial and will require time,
it may in the long run lead to a simpler physical understanding
of the coronal expansion than subscribed to in the recent past.

5. Summary

The relationship between coronal holes and the interplanetary stream-sector structure has been enormously clarified by
the observations of both during the Skylab epoch. The general
association of holes with high-speed streams is confirmed by
the comparisons described above. The recognition of polar
extensions as coronal holes capable of producing high speed
streams near the solar equator leads to a nearly one-to-one
relationship between hole-like sources and interplanetary
streams. The magnetic sector structure is clearly associated
with the general pattern of coronal holes of a given polarity;
this pattern may drift relative to the individual coronal
holes on the sun and lead to interplanetary sectors that recur
with a period different from the 27-day recurrence period of
streams. The development of a very stable coronal structure
in 1974 led to a pair of recurrent geomagnetic disturbances
that persisted with little change for two years. These storms
were correlated with a stable two-sector, two-stream interplanetary structure with speed variations of large amplitude.
Their coronal sources were a pair of polar extensions of opposite polarity and large area.

These observations can be unified into a rather simple
phenomenological description of the three-dimensional structure of the corona and interplanetary space by a simple
assumption relating coronal holes to open magnetic regions in
the corona. For the Skylab epoch this description implies a
dominant influence by a dipole component of the solar magnetic
field, with interplanetary sector boundaries related to the
tilting, warping and distortion of a neutral sheet encircling
the sun within $\sim 30^{\circ}$ of its equator. High-speed solar wind
emanates from the polar regions of the sun and from any large
coronal holes; low speed wind occurs in a "belt" $\sim 50^{\circ}$ wide,

centered on the neutral sheet. Thus the 700-750 km sec^{-1} solar wind speeds commonly observed in 1973 and 1974 in the ecliptic plane, may be more typical of global solar wind conditions than that observed within the low-speed belt during the intervening decade. The folly of basing our knowledge of the coronal expansion on measurements made solely within the ecliptic plane, should be obvious.

This clarification of coronal and interplanetery phenomenology has not yet been matched by a similar advance in our physical understanding of the processes giving rise to the solar wind. The situation most amenable to quantitative models, the identity of coronal holes with open magnetic structures, may not be consistent with observed magnetic fluxes in coronal holes. The high expansion speeds observed in 1973 and 1974 imply a significantly larger energy flux in the wind than previously measured; mapping of this flux to the limited areas covered by coronal holes at the base of the corona implies an energy flux of $\sim 5 \times 10^5$ ergs cm^{-2} sec^{-1} to supply the solar wind. A flux of this size implies that the wind is the dominant energy loss from open regions in the corona and is a severe requirement for any models of the expanding corona.

Acknowledgements

The author has benefitted enormously from the unselfish sharing of ideas, data, and enthusiasm displayed by many participants in the Coronal Hole Workshop. A special debt must be acknowledged to those who contributed unpublished data and figures to this article; specific credits are found in the figure captions displaying this material. Comments on the manuscript by T. E. Holzer and J. B. Zirker were of great value. The author is also grateful to J. Bazil who drew the figures and to B. Kirwin for typing the manuscript.

References from <u>Coronal Holes and High Speed Wind Streams</u>, edited by Jack B. Zirker, Colorado Associated University Press, Boulder, Colorado, 1977:

1. Bohlin, J. D., An observational definition of coronal holes, pp. 27-69.

2. Kopp, R. A., and Orrall, F. Q., Models of coronal holes above the transition region, pp. 179-224.

3. Krieger, A. S., Temporal behavior of coronal holes, pp. 71-102.

4. Levine, R. H., Large-scale solar magnetic fields and

coronal holes, pp. 103-143.

5.   Svalgaard, L., Geomagnetic activity, dependence on
     solar wind parameters, pp. 371-441.

6.   Zirker, J. B., Coronal holes - an overview, pp. 1-26.

General References

7.   Broun, J. A.   1858, Philos. Magazine, 16, 81.

8.   Chapman, S. and Bartels, J.   1940, Geomagnetism   (The
     Clarendon Press, Oxford).

9.   Bartels, J.   1932, Terr. Magn. and Atmos. Elect., 37, 1.

10. Mustel', E. R.   1964, Space Sci. Rev., 3, 137.

11. _____.   1944, Dokl. Acad. Nauk. SSR, 42, 117.

12. _____.   1967, Soviet Astron., AJ, 10, 899.

13. Allen, C. W.   1944, Mon. Not. R. Astron. Soc., 104,
     13.

14. _____.   1964, Planet. Space Sci., 12, 487.

15. Pecker, J. -C., and Roberts, W. O.   1955, J. Geophys.
     Res., 60, 33.

16. Saemundsson, Th.   1962, Mon. Not. Roy. Astron. Soc., 123,
     299.

17. Roelof, E. C. 1974, Solar Wind Three, ed. C. T. Russell
     (UCLA).

18. Gulbrandsen, A.   1975, Planet. Space Sci., 23, 143.

19. Parker, E. N.   1958, Astrophys. J., 128, 664.

20. Snyder, C. W., and Neugebauer, Marcia   1963, J. Geophys.
     Res., 68, 6361.

21. Neugebauer, Marcia, and Snyder, C. W.   1966, J. Geophys.
     Res., 71, 4469.

22. Bame, S. J., Asbridge, J. R., Feldman, W. C., and Gosling,
     J. T.   1976, Astrophys. J., 207, 977.

23. Wilcox, J. M.   1968, Space Sci. Rev., 8, 258.

24. Wilcox, J. M., and Ness, N. F. 1965, J. Geophys. Res.,
    70, 5793.

25. Ness, N. F., and Wilcox, J. M. 1967, Astrophys. J., 143,
    23.

26. Snyder, C. W., and Neugebauer, Marcia, The Solar Wind,
    ed. R. J. Mackin, Jr., and Marcia Neugebauer
    (Pergamon Press, New York).

27. Hundhausen, A. J. 1972, Coronal Expansion and Solar Wind
    (Springer-Verlag, New York).

28. Billings, D. E., and Roberts, W. O. 1964, Astrophysica
    Norvegica 9, 147.

29. Pneuman, G. W., and Kopp, R. A. 1971, Solar Phys., 18,
    258.

30. Waldmeier, M. 1957, Die Sonnenkorona, Vol. 2 (Verlag
    Birkhäuser, Basel).

31. _____. 1975, Solar Phys., 40, 351.

32. Newkirk, G. A. 1967, Ann. Rev. Astron. Astrophys., 5,
    213.

33. Altschuler, M. D., Trotter, D. E., and Orrall, F. Q.
    1972, Solar Phys., 26, 354.

34. Krieger, A. S., Timothy, A. F., and Roelof, E. C. 1973,
    Solar Phys., 29, 505.

35. Bell, B., and Noci, G. 1976, J. Geophys. Res., 81, 4508.

36. Neupert, W. M., and Pizzo, V. A. 1974, J. Geophys. Res.,
    79, 3701.

37. Hansen, R. T., Hansen, S. F., and Sawyer, C. 1976,
    Planet Space Sci., 24, 381.

38. Wagner, W. J. 1975, Astrophys. J., 198, L141.

39. McIntosh, P. S., Bull. Amer. Astron. Soc., 8, 325 (ab-
    stract).

40. Hundhausen, A. J., Hansen, R. T., Hansen, S. F., Feldman,
    W. C., Asbridge, J. R., and Bame, S. J. 1976,
    AGU International Symposium on Solar-Terrestrial
    Physics.

41. Nolte, J. T., Krieger, A. S., Timothy, A. F., Gold, R. E.,
    Roelof, E. C., Vaiana, G., Lazarus, A. J., Sullivan,
    J. D., and McIntosh, P. S. 1976, Solar Phys., 46,
    303.

42. Bohlin, J. D., and Rubenstein, D. M. 1975, Report UAG
    51 (World Data Center A for Solar-Terrestrial Phy-
    sics, NOAA, Boulder, Colo.).

43. Wagner, W. J. 1976, Astrophys. J., 206, 583.

44. Sheeley, Jr., N. R., Harvey, J. W., and Feldman, W. C.
    1976, Solar Phys., 49, 271.

45. Svalgaard, L. 1976, Stanford University Institute for
    Plasma Research Report No. 648.

46. _____. 1975, J. Geophys. Res., 80, 2717.

47. Gosling, J. T., Asbridge, J. R., Bame, S. J., and Feld-
    man, W. C. 1976, J. Geophys. Res., 81, 5061.

48. Hundhausen, A. J., Bame, S. J., Asbridge, J. R., and
    Sydoriak, S. J. 1970, J. Geophys. Res., 75, 4643.

49. Brandt, J. C. 1967, Planet. Space Sci., 15, 941.

50. Nolte, J. T., and Roelof, E. C. 1973, Solar Phys., 33,
    241.

51. Poland, A. 1978, Solar Phys. in press.

52. Rosenberg, R. L., and Coleman, P. J. 1969, J. Geophys.
    Res., 74, 5611.

53. Rosenberg, R. L. 1970, Solar Phys., 15, 72.

54. Sawyer, C. 1974, Geophys. Res. Lett., 1, 7.

55. Sime, G. D. 1976, Ph.D. thesis (University of Califor-
    nia, San Diego).

56. Hundhausen, A. J. 1973, J. Geophys. Res., 78, 1528.

57. Hundhausen, A. J., and Pizzo, V. 1977, submitted to J.
    Geophys. Res.

58. Svalgaard, L., and Wilcox, J. M. 1975, Solar Phys., 41,
    461.

59. Newkirk, G. A. 1972, Solar Wind, ed. C. P. Sonett, P. J. Coleman, Jr., and J. M. Wilcox, NASA SP-308 (Washington).

60. Schatten, K. H. 1972, Solar Wind, ed. C. P. Sonnett, P. J. Coleman, Jr., and J. M. Wilcox, NASA SP-308 (Washington).

61. Ness, N. F., Searce, C. S., and Seek, J. B. 1964, J. Geophys. Res., 69, 3531.

62. Suess, S. T., Richter, A. K., Winge, C. R., and Nerney, S. F. 1976, submitted to Astrophys. J.

63. Pizzo, V. 1977, Ph.D. Thesis (University of Colorado).

64. Endler, F. 1971, Ph.D. Thesis (Göttingen University).

65. Pneuman, G. W. 1973, Solar Phys., 28, 247.

66. Altschuler, M. D., Newkirk, G. A., and Trotter, D. E. 1971, Solar Magnetic Fields, IAU Symposium No. 43.

67. Stix, M. 1977, Astron. Astrophys., 59, 73.

68. Wilcox, J. M. 1970, Intercorrelated Satellite Observations Related to Solar Events, ed. V. Manno, and and D. E. Page, (Reidel, Dordrecht).

69. _____. 1971, Physics of the Solar Corona, ed. Macris, (Reidel, Dordrecht).

70. _____. 1971, Comments on Astrophys. and Space Phys., 3, 133.

71. Svalgaard, L., Wilcox, J. M., and Duvall, T. L. 1974, Solar Phys., 37, 157.

72. Schulz, M. 1973, Astrophys. and Space Sci., 24, 371.

73. Alfvén, H. 1977, Rev. Geophys. Space Phys., 15, 271.

74. Svalgaard, L., Wilcox, J. M., Scherer, P. H., and Howard, R. 1975, Solar Phys., 45, 83.

75. Hansen, S. F., and Svalgaard, L. 1977, Topical Conference on Solar and Interplanetary Physics, (Tucson, Arizona).

76. Svalgaard, L., and Wilcox, J. M. 1976, Nature, 262, 766.

77.   Smith, E. J., Tsurutani, B. T., and Rosenberg, R. L.
          1976, (abstract) EOS, 57, 997.

78.   Bame, S. J., Asbridge, J. R., Feldman, W. C., and Gosling,
          J. T. 1977, J. Geophys. Res., 82, 1487.

79.   Cortie, A. L. 1912, Mon. Not. Roy. Astron. Soc., 73,
          52.

80.   Brandt, J. C., Harrington, R. C., and Roosen, R. G.
          1975, Astrophys. J., 196, 877.

81.   Bame, S. J., Asbridge, J. R., Feldman, W. C., Felthauser,
          H. E., and Gosling, J. T. 1977, J. Geophys. Res.,
          82, 173.

82.   Schwenn, R., Montgomery, M. D., Rosenbauer, H., Miggen-
          rieder, M., Mühlhäuser, K. H., Bame, S. J., Feldman,
          W. C., and Hansen, R. T. 1978, J. Geophys. Res.,
          83, 1011.

83.   Rhodes, Jr., E. H., and Smith, E. J. 1976, J. Geophys.
          Res., 81, 5833.

84.   Broussard, R. M., Sheeley, N. R., Jr., Tousey, R., and
          Underwood, J. H. 1978, to be published in Solar
          Phys.

85.   Gosling, J. T., Asbridge, J. R., and Bame, S. J. 1977,
          J. Geophys. Res., 82, 3311.

86.   Sheeley, N. R., Asbridge, J. R., Bame, S. J., and
          Harvey, J. W. 1977, Solar Phys., 52, 485.

87.   _____. 1976, J. Geophys. Res., 81, 3462.

88.   Wilcox, J. M. 1972, Comments on Astrophys. and Space
          Phys., 4, 141.

89.   Hundhausen, A. J., Bame, S. J., and Montgomery, M. D.
          1971, J. Geophys. Res., 76, 5145.

90.   Stix, M. 1974, Astron. Astrophys., 37, 121.

91.   Howard, R. 1976, Astrophys. J., 210, L159.

92.   Durney, B. R., and Pneuman, G. W. 1975, Solar Phys.,
          40, 461.

93.   Ness, N. F. 1970, Space Sci. Rev., 11, 459.

94. Munro, R., and Jackson, B. 1977, Astrophys. J., 213, 874.

95. Burlaga, L. F., Behannon, K. H., Feldman, W. C., Hansen, S. F., and Pneuman, G. W. 1977, preprint.

96. Howard, R. 1972, Solar Wind, ed. C. P. Sonett, P. J. Coleman, Jr., and J. M. Wilcox, NASA SP-308 (Washington).

97. Feldman, W. C., Asbridge, J. R., Bame, S. J., and Gosling, J. T. 1976, J. Geophys. Res., 81, 5054.

98. Kopp, R. A., and Holzer, T. E. 1977, Solar Phys., 49, 43.

99. Steinolfsen, R. S., and Tandberg-Hanssen, E. 1976, NASA/MSFC Space Sciences Preprint #76-4.

100. Wang, Y-C., and Chien, T. H. 1977, submitted to Astrophys. J.

101. Pneuman, G. W. 1976, Physics of Solar Planetary Environments, ed. D. J. Williams, (American Geophysical Union).

102. Schatten, K. H., Wilcox, J. M., and Ness, N. F. 1969, Solar Phys., 6, 442.

103. Altschuler, M. D., and Newkirk, G. A. 1969, Solar Phys., 9, 131.

104. Pneuman, G. W. 1966, Astrophys. J., 145, 242.

105. _____. 1976, J. Geophys. Res., 81, 5049.

106. Barnes, A. 1975, Rev. Geophys. Space Phys., 13, 1049.

107. Belcher, J. W. 1971, Astrophys. J., 168, 509.

108. Hollweg, J. V. 1975, Rev. Geophys. Space Phys., 13, 263.

109. Holzer, T. E. 1976, Physics of Solar Planetary Environments, ed. D. J. Williams, (American Geophysical Union).

110. _____. 1977, J. Geophys. Res., 82, 23.

5

Seismic Sounding of the Sun

Henry A. Hill

In this review we present recent observational data
which suggest the existence of global oscillations of the
sun that are of sufficient magnitude as to be observable.
Should these findings be correct, then a new observational
"tool" would be available to astrophysicists which would
permit seismic sounding of the sun as a valuable probe of
solar structure. The detection of such oscillations by an
observational method which permits a precisely reproducible
definition of an edge on the solar disk is discussed in
relation to other types of observational techniques, techniques which have been considered to be capable of the
detection of such oscillations, but which have failed to do
so. Possible avenues to the resolution of such observational
discrepancies are considered in the context of traditional
pulsation theory, major categories of observational methods
by which solar oscillations might be detectable, and present
evidence for the existence of solar oscillations. The
implications of "seismic sounding" of the sun through the
study of global oscillations are considered for solar and
stellar modeling. The relevance of what may be the first
important results of seismic sounding is examined, and the
directions of current studies are discussed.

1. Introduction

In the fall of 1973, while studying the oblateness of
the sun, we obtained what we believe to be the first evidence
suggesting that the entire body of the sun is oscillating.
The simplest modes of oscillation are spherical expansions
and contractions of the sun, analogous to the expansion and
contraction of a balloon as the amount of air inside it is
changed; more complex oscillatory motions are also possible
in which the sun expands in one or more regions while simultaneously contracting in others. The primary observational
technique used in this program was the measurement of the

separation of two diametrically opposite edges on the solar
disk as a function of time. One of the more formidable problems in such a measurement is that of locating the edge of the
sun, an entity relatively simple to conceptualize but difficult to define observationally. The difficulty arises from
the fact that the solar edge is more an abstraction than a
physical entity, much as is the outer "edge" of the earth's
atmosphere. The solar edge, as defined by a certain technique,
will depend upon a number of factors including the intensity
distribution on the solar disk and the refractive effects of
the earth's atmosphere, which tend to degrade the sensitivity
of the measurements. For such reasons the above determination
of the separation between two edges oppositely positioned on
the solar diameter is frequently referred to as a measurement
of the "apparent" diameter. An indication of the complex
nature of the edge definition problem is given by the fact
that nearly four years were required in order to improve
existing edge definition techniques to a degree sufficient to
yield consistently reproducible results. This work resulted
in the development of a method which improved by a factor of
100 to 1000 the accuracy to which we were able to define an
edge compared to that which was previously available.

The first results of our 1973 study of solar oblateness
using the new edge definition technique were reported in 1974
(1), presenting solar diameter measurements which exhibited
periodic fluctuations on the time scale of approximately half
an hour. As this structure was not at all typical of random
noise ((2), Fig. 10) we later theorized that this fluctuation
might represent measurable evidence of global solar oscillations (3). These presumed oscillations had periods between
approximately 10 min and one hour; their amplitudes were
quite small, being on the order of a few parts per million
of the solar radius. If such oscillations do, in fact, exist,
they will provide a unifying link between solar physics and
various other areas of astrophysics as well as a potentially
important tool for improving the modeling of stellar interiors
and atmospheres. The oscillations will offer a new diagnostic
technique for probing the interior of the sun in much the same
manner as that used by seismologists to study the earth's interior. They may also permit a quantitative examination of
the current solar model with a degree of accuracy that has
rarely been achievable in astrophysics.

One might ask why the discovery of solar oscillations of
global dimensions was not made at an earlier date in history.
The answer to this question may lie partly in the fact that
there was no method by which the solar edge could be defined
precisely enough to permit a sufficient degree of accuracy in
the measurement of the solar diameter as to render the

oscillations observable. Additionally, given the prevailing
belief that the sun was essentially stable, there was little
reason to seek such an interpretation of observational data
from other sources. The technique developed at SCLERA (4)
provided the necessary tool for detection of the oscillations.

The concept of global solar oscillations and of seismic
sounding of the solar interior is not a new one; it surely
has received in the past at least brief consideration by some
astronomers (and perhaps by a few nonastronomers). However,
the "Achilles' heel" of the idea has always been the absence
of a plausible, theoretical physical excitation process capable of producing solar oscillations of observable magnitude.
Consequently, a strong consensus has evolved over the years
in favor of the representation of the sun as a stable, nonpulsating star. Contradicting this view are recent findings
such as those of SCLERA and others (see discussion by Eddy in
this volume) that the sun may not be as stable as has been
traditionally believed.

Objections to the concept that the sun is an oscillating
body have been based upon several grounds. From an observational perspective, the sun does not appear to be a member of
any of the known classes of variable stars (5). From the stand-
point of traditional theory, it has also been generally concluded that none of the destabilizing mechanisms which are
known to lead to stellar pulsation are operative in the sun.
No one seriously questioned the traditional theories of a
stable sun until the recognition of the solar neutrino paradox (see discussion by Davis in this volume) initiated a
search for applicable destabilizing mechanisms in the sun
(6-10). In addition to the lack of a theoretical destabilizing
mechanism, it should also be noted that no one has postulated
processes which could serve to limit the magnitude of a solar
pulsation. If the sun is in fact oscillating, the magnitude
of the oscillations must necessarily be relatively small or
they would surely have been detected some time ago. At such
a small amplitude of oscillation, it is likely that none of
the standard limiting processes would be operative. Finally,
there are certain less tangible human factors which have
traditionally served to resist the expression and implementation of those new ideas which threaten established beliefs.
A notable example may be found in Auwers' treatment of solar
observations at the turn of the century (11). Auwers assumed
that any observed changes in the shape of the sun were produced by the methods and/or perceptions of each individual
observer; hence, his practice was to adjust the deviating
results of certain observers by the insertion of appropriate
individualized weighting functions. There is also a certain
measure of psychological comfort to be derived from a

conceptualization of our sun as a stable star and, furthermore, treatment of the sun as an oscillating body adds new
theoretical complexities, seemingly running counter to the
practice of applying the principle of Ockham's razor to scientific theory.

Despite the foregoing objections, some observational
evidence has been derived over the past fifteen years which
may lend support to the concept of an oscillating sun. While
the sun has not generally been regarded as a pulsating star,
the coherence of certain types of motions over large areas
of the sun's surface as observed by Musman and Rust (12)
prompted Wolff (13) to suggest that the entire sun might be
pulsating in high order nonradial modes. Preliminary calculations by Wolff (13), whose results have been essentially
reproduced by the detailed calculations of Ando and Osaki (14),
show that such modes are strongly excited by opacity variations (the κ- mechanism of Baker and Kippenhahn (15)) and the
superadiabaticity of the temperature gradient (16) in the
outer subphotospheric solar layers. Wolff (17) further suggested that giant solar flares might excite such oscillations,
just as earthquakes can set the entire planet "ringing".

In recent years, significant progress has been achieved
in both observational and theoretical work on solar oscillations. Observational evidence is now available which strongly supports the contention that the sun is pulsating. These
diverse observations are reviewed in §2. The observations
fall naturally into three categories which may be classified
as measurements of surface velocity in the direction of the
observer, brightness variations, and of the diameter of the
solar disk. Subsequent intercomparisons between results of
these categories of study have led to various discrepancies:
from diameter measurements a velocity is predicted but not
observed; from velocity measurements, brightness changes are
predicted but not observed; and from diameter measurements,
brightness changes are predicted but likewise are not
observed. These discrepancies appear to arise from fundamental inadequacies in the current theoretical treatment of such
phenomena and thus they have taken on the character of paradoxes. It should also be noted that since paradoxes arise
from each of the intercomparisons, no single observational
category can be considered the culprit. In order to illuminate the fundamental nature of these paradoxes, §3 describes
the types of measurements used in each of the three categories
and discusses the methods which have been used for their
intercomparison.

For reasons discussed in §3 concerning the methods of
intercomparing observations, a resolution of these paradoxes

is likely to be achieved only if the sensitivity of the various observational methods to oscillations is carefully reexamined. This sensitivity is inextricably connected to the manifestation of solar oscillations in the sun's atmosphere where they are observed. The geometric aspects of the sensitivity of the observations are considered in §4 while the physics of the oscillations in the solar atmosphere are analyzed in §5. It is apparent from this discussion that for the first time, the outer boundary conditions used in stellar pulsation theory are no longer relegated to secondary roles.

The first results of seismic sounding of the sun, which already comprise an impressive list, have been achieved by the careful execution of the types of calculations discussed in §4 and §5. These results and their implications are examined in §6.

The findings discussed in §6 may only be the harbingers of a new body of solar seismological information to be added to our knowledge of solar physics. If solar oscillations are not subsequently shown to be artifacts, the prospects for solar and stellar seismology are both promising and impressive. Some of the more interesting possibilities are discussed in §7. These concern the possible modification of stellar envelope models to reflect the presence of oscillations, a limiting mechanism for solar oscillations, the study of spectral line theory, tests of nonlinear pulsation theory and associated mechanisms in the solar atmosphere, a method for detailed examination of the boundary conditions used in pulsation theory, a means by which to study the solar interior and observation of internal solar rotation. The success realized to date has been achieved through an "on line" interaction of observations and theory and the degree of future success will depend strongly upon its continuation and the development of additional associations of this nature.

2. Observational Evidence for
Global Oscillations

Although the sun has traditionally been considered a nonpulsating star, much evidence obtained within the last decade has cast doubt on this assumption of solar stability. However, in the language of a mathematician concerning the necessary and sufficient conditions in the proof of a theorem, this evidence meets primarily only the necessary condition of the proof that oscillations are indeed global. At this time the evidence to meet the sufficient condition in the proof is still lacking in varying degrees for the different types and modes of global oscillations (18). This insufficiency stems, in some cases, from possible alternative interpretations of

the observations and, in other cases, from the development of
very clear discrepancies which hinge on certain interpretations of the observations. The following subsections present
the observational evidence for global oscillations; the paradoxes generated by comparing different kinds of data and possible pathways to their resolution will be discussed in later
sections.

The ground-based observations relating to solar oscillations fall naturally into three major categories according to
the reported periods of oscillation. The oscillation periods,
Π, are: (1) $\Pi \sim 5$ min; (2) 5 min $< \Pi \lesssim 1$ hr; and (3) $\Pi > 1$
hr. A fourth category includes oscillations with observed
periods of $\Pi < 5$ min (19); it has been suggested that these
oscillations may be important in the generation of microtur-
bulence in the solar atmosphere. At the other extreme end of
the spectrum, Wolff (20) suggests that there may be a causal
relation between global oscillations and the long-term structure of the sunspot cycle. Although these interpretations
both address very fundamental questions in solar physics, the
following discussions are restricted to the first three
classes of oscillations listed above.

There have been several extensive reviews on motions in
the solar atmosphere over the years.(e.g., 21, 22) The following sections will review only that work not covered in
these previous reports. While this restriction will limit
the scope of the general discussion, it will place little
restriction upon the discussion of global oscillations since
most of the critical evidence for their existence has only
been obtained within the last few years.

2.1 Five-Minute Oscillations

Those solar oscillations having a period of five minutes
are the best documented. The initial evidence for their
existence was first reported in 1960 (23). They may be observed by monitoring brightness changes of the continuum
radiation (i.e., the radiation not associated with spectral
lines), or by observing intensity changes in the spectral
lines (which could be due to temperature changes in the atmosphere produced by the oscillations), and by studying surface
velocities using the Doppler shifts of spectral lines.

This solar oscillation typically manifests itself as a
small scale (less than 5000 km) velocity field in the solar
photosphere and low chromosphere. The motions are predominantly radial with a period of about 300 sec. The lifetime
of a given oscillation, as measured by the decay of the

velocity-time autocorrelation function, is only about two
periods of oscillation.

The five-minute oscillations have often been analyzed
theoretically as a purely atmospheric phenomenon (24-30).
These analyses have treated the photosphere as a rigid boundary or as a layer with an imposed turbulent boundary condition. Within the atmosphere both acoustic waves and gravity
waves have comparable frequencies although the spatial characteristics are distinct; because most observational work has
dealt primarily with frequency characteristics of the five-
minute oscillation, some controversy over the actual nature of
the waves has resulted. Interpretation of these particular
oscillations as acoustic modes is presented in the first four
of those papers mentioned above while the last three have
treated the oscillations as gravity modes. Frazier (31) has
offered persuasive evidence that at least a major portion of
the oscillating power of the five-minute oscillations is in
the form of acoustic modes.

A different class of models based upon trapping of acoustic waves below the photosphere has been considered in some
detail by Ulrich (32), Leibacher and Stein (33), and Wolff
(13). These models offer an explanation for the existence of
a wave motion in a layer of the quiet solar atmosphere in a
frequency band where waves are essentially nonprogressive.
Similar results have been obtained by Ando and Osaki (14), who
treated these oscillations as global nonradial modes.

The "modal" character of these overstable subphotospheric
oscillations is theoretically characterized as a concentration of power along ridges in a figure where the axes are k
and ω, k representing the horizontal wavenumber and ω the
eigenfrequency. Despite extensive work, this "modal" character has only recently become evident in observations, primarily because of the large accumulation of data necessary for
the computation of k,ω spectra with adequate statistical
stability.

The first clear observational resolution of the power
into ridges on the k-ω plane is found in the work by Deubner
(34). Figure 2.1 is taken from a more recent work by Deubner
(35); Rhodes, Ulrich and Simon (36) have corroborated and more
precisely defined this structure, and on the basis of these
results have concluded that the five-minute oscillations
clearly represent nonradial p mode oscillations in the solar
envelope.

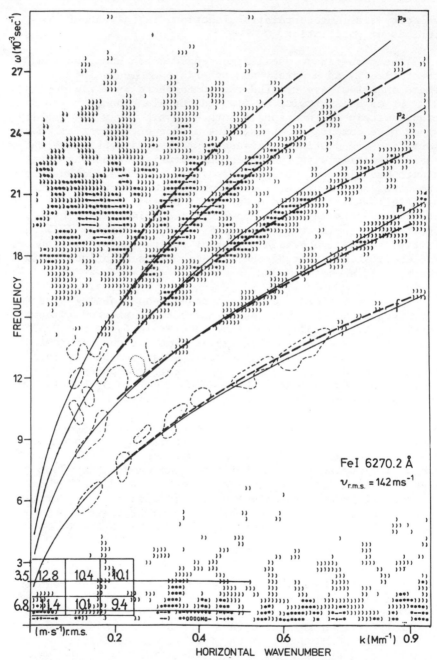

Fig. 2.1. Power spectrum of velocity fluctuations (Doppler shifts of the Fe I 6270.2 spectral line) as a function of the angular frequency ω, and the horizontal wave number, k, which were obtained by Deubner (35). The solid lines are the places where Ando and Osaki (14) predicted power to be concentrated.

2.2   Oscillations with Periods Between Five Minutes and One Hour

The discovery of solar oscillations with periods between five minutes and one hour would be significant, since these oscillations might represent low order p modes, i.e. pressure modes. Deubner (37) and Fossat and Ricort (38) reported the occasional occurrence of a 40 min oscillation possibly related to the incidence of solar flares. Subsequently Fossat (39) described an oscillation with a 10 min period whose occasional presence was tentatively confirmed by Livingston (40). These observations all utilized Doppler techniques.

However, it was not until the introduction of the new observational techniques used at SCLERA that the first evidence was obtained which indicated that long, large-scale oscillations with periods between five minutes and one hour could be detected (3). Observations at SCLERA using these techniques, which will be explained in §3, have continued to indicate the existence of oscillations of this character (41-43). During this period, conflicting results obtained by established techniques of Doppler and intensity measurements have also been reported (44-51). Clearly, from an observational point of view, the primary points under consideration in this field today are contained within the following questions. Do the observations at SCLERA indicate fluctuations that are repeatable in nature, or do they originate in statistical fluctuations in a broad-band noise source? If they are repeatable, are they of solar origin, or are they generated within the instrument or the earth's atmosphere? If they are both repeatable and solar in origin, are their frequencies well defined and compatible with expected normal mode oscillations in the sun? These questions are briefly addressed in the following paragraphs.

2.2.1 Statistical Tests   The statistical significance of the evidence for the existence of oscillations with periods between 5 min and one hour reported by Hill, Stebbins and Brown (41) and Brown, Stebbins and Hill (42,52) has been questioned by other workers (45). The results of several statistical tests performed by Brown, Stebbins and Hill (42, 52) are briefly summarized below.

A method for the direct testing of the reproducibility of the measured daily power levels may be derived from the work by Groth (53). Groth demonstrated that the expected measured power of the power spectrum of a nonvarying signal mixed with random noise could be represented as

$$\overline{P} = 1 + P_s, \qquad (2.1)$$

while the standard deviation of this measured power was

$$\sigma = (1 + 2P_s)^{\frac{1}{2}} . \tag{2.2}$$

All of the above quantities are measured in units of the mean noise power; P_s is the signal power. Both σ and $\overline{P}$ were estimated from the daily power spectrum of eleven data sets each having an average length of 7.1 hours and obtained on eleven different days between October 9 and November 6, 1975. In examining a band of frequencies between 0.2 and 1.0 mHz in this way, Brown, Stebbins and Hill (52) found an estimate of $\sigma/\overline{P}$ equal to 0.86 ± 0.06. This value is significantly different from a pure noise source. Furthermore, the estimate of 0.86 implies that $P_s \simeq 1$, indicating that the ratio of non-varying signal power to random noise power is approximately one.

The second test by Brown, Stebbins and Hill (42,52) examines the above data sets more directly. Intuitively, one might expect that if most of the signal power resides in the peaks in the power spectrum, these peaks should be inherently repeatable. To test this idea, two independent average power spectra were formed containing the first five and the last six of the eleven above-mentioned spectra. The numbers of peaks coinciding in the two averages were noted. In the case of random noise, coincidences were to be expected 1/3 of the time if every peak was counted, and less often if only the larger (and correspondingly rarer) peaks were included in the count. They found that if peaks examined included only those at least 20 (milli-arcsec)^2 larger than their surroundings, 10 out of 15 peaks were coincident. The probability of obtaining this result from a pure noise source, calculated by a binomial distribution, is 5×10^{-4}. This corresponds to a deviation of 3.5 σ on a normal probability distribution. When every peak was counted, the deviation from random behavior was equal to 2.5 σ, reinforcing the assumption that the larger peaks were more reproducible than the smaller peaks.

A test more directly applicable to long period oscillations was reported by Brown, Stebbins and Hill (52); it examined the phase of an oscillation as a function of time. The absence of a frequency which will fit the phase data within the observational limits indicates that the power at that frequency is probably not due to global oscillations. On the other hand, the discovery of a suitable frequency gives statistical significance to the classification of the power at that frequency as being representative of a global oscillation of solar origin, as noted below. A much more accurate value for the frequency of oscillation is also obtained.

Brown, Stebbins and Hill (52) analyzed the phase data of the
two previously identified long period peaks in the power spectra having approximate periods of 68 min and 45 min, and were
able to find frequencies for both peaks which satisfied the
phase data within observational errors over a period of 26
days. The probability of random occurrence was 0.16 for the
45 min peak and 0.54 for the 68 min peak.

A similar result has been found by Hill and Caudell (43)
upon reanalysis of the 1973 solar oblateness observations of
Hill and Stebbins (54). The power spectrum analysis exhibited
six peaks below the Nyquist frequency of the solar oblateness
data set, five appearing at frequencies close to the first five
listed by Brown, Hill and Stebbins (42, 52). Furthermore, the
six oscillations were found to be phase coherent over a thirteen
day period. In addition, phase coherence analyses of the
oscillations having periods of 68 min and 45 min yielded good
agreement between the two observational sets of data obtained
two years apart.

These results substantially conflict with the assumption
that the observed low frequency power is due entirely to
noise, but are consistent with the presence of global oscillations. It is probable that a large number of the peaks in
this frequency range represent real signal power. The issue
of the observed oscillations in the apparent solar diameter
thus centers around the identification of the source of these
repeatable signals.

2.2.2 Possible Origin of Observed Oscillations   Several properties of the observed oscillations indicate that they
must be solar in origin. One of these is the apparent step in
the power level at about 5.5 mHz for the averaged power spectrum of Brown, Stebbins and Hill (42, 52). This is near the
acoustic cut-off frequency of 5.4 mHz in the photosphere, a
frequency so characteristically solar in origin that its
generation by the telescope or the earth's atmosphere is
doubtful.

A second property is the phase coherence reported by
Brown, Stebbins and Hill (52) and by Hill and Caudell (43). A
large majority of systematic errors in any telescope system
are related to the pointing of the device and consequently to
the time of day. Problems associated with differential or
terrain-influenced atmospheric refraction fall into this category. Such difficulties should all manifest themselves at
periods of an exact integral number of cycles per day. However, the periods obtained by the phase analysis are strongly
inconsistent with any source of variation which changes in a

diurnal fashion, or which is phase-locked to the start-up of the telescope system.

A third property recently discovered by Hill and Caudell (43) presents the strongest evidence yet available in support of the solar origin interpretation. The solar oblateness observations used in this analysis are unique in that diameter measurements were made essentially simultaneously at two different detector geometries in the technique used to define the solar edge (see discussion in §3.1.3 and Appendix A). Due to inherent properties of the definition, a determination of high dependence of the observed oscillatory power upon the detector geometry characterized by the range of integration or scan amplitude in equation (3.1) strongly indicates the correctness of a solar interpretation of the scan amplitude-dependent portion of that power. The analysis of Hill and Caudell (43) demonstrated a strong scan amplitude dependence in the power spectrum, implying that a significant fraction of the observed power must in fact be solar in origin. These new results are, in addition, quite significant in relation to the boundary condition problem discussed in §4 and §6.1.

2.2.3 Comparison of Observed Periods with Theoretical Eigenfrequencies   Power spectra of long time series of the apparent solar diameter as measured at SCLERA have been previously discussed (42,52). The comparison of these observations with the theoretically predicted frequencies of oscillation, i.e. eigenfrequencies, has to date assumed that the excited modes were p modes and that they had a very large horizontal scale (the principle order number, ℓ, of the spherical harmonic representing the eigenfunction was assumed to be $\simeq$ zero). Good general agreement has been demonstrated in these analyses (55-59). An example of this agreement is contained in Table 2.1, where the observational results are from (52) and the $\ell = 0$ theoretical results are from Christensen-Dalsgaard and Gough (55).

The good agreement that can be obtained by assuming $\ell \simeq 0$ may be one of the reasons that only p modes of small ℓ have received such attention. The rationale for exclusion of p modes of larger ℓ and of g modes from consideration is unclear, particularly in view of the evidence given by Hill and Caudell (43) which strongly suggests that g modes are in fact excited and also since agreement similar to that in Table 2.1 can be obtained when consideration is not restricted to $\ell = 0$. Such a comparison is examined below for completeness.

The relevant question in this case deals with the description of observations in terms of excitation of oscillations not restricted to low order ℓ. The eigenfrequencies

Table 2.1

Observed and Predicted Oscillation Periods

Observed Period[a] minutes	Predicted Period[b] (for $\ell = 0$ modes) minutes
66.25	62.22
44.66	
	41.98
39.00	
32.1	32.32
28.7	
24.8	26.00
21.0	21.51
19.5	18.33
	15.95
13.3	14.13
12.1	12.64
11.4	11.54
10.7	10.60
9.9	9.81
9.3	9.12
8.5	8.52
7.90	7.94
7.54	7.53
7.18	7.12
6.74	6.75
6.28	6.41
6.07	6.10

[a]Brown, Stebbins and Hill (52)

[b]Christensen-Dalsgaard and Gough (55)

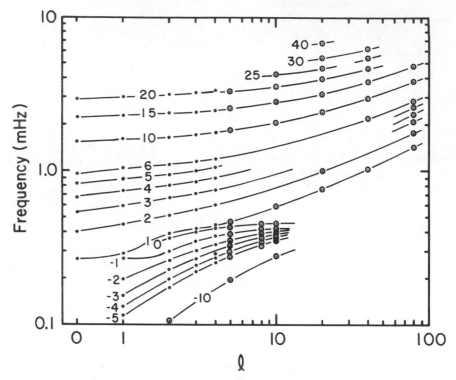

Fig. 2.2. Theoretical eigenfrequency spectrum for a standard
solar model as a function of ℓ, the principle order
number of the spherical harmonic associated with
a particular mode. The curves are designated by a
parameter k; modes with positive and negative values
of k are p modes and g modes respectively, while the
k = 0 modes are referred to as fundamental modes.
The theoretical results shown as dots and circles
are from Iben and Mahaffy (59) and Wolff (60) respectively.

that have been obtained in various calculations are shown as
a function of ℓ in Figure 2.2.   Inspection of the figure
demonstrates that the observed peak in power at 66.25 min
(Table 2.1) is clearly a good candidate for the possible contribution of p and g modes. It may also be seen that if g
modes, higher order in ℓ, were excited as noted by Hill and
Caudell (43) the peak in power at 44.66 min (Table 2.1) could
also represent a combination of p and g modes. However, Figure 2.2 shows that it is not possible for the spectra with
periods shorter than 45.5 min to contain contributions from g
modes (assuming that the solar interior is reasonably similar
to a standard solar model). This observation is important in
the comparison of velocity type measurements with SCLERA observations.

In dealing with periods $\lesssim$ 40 min and the probable presence of p modes alone, the theoretical power spectrum that
would be expected can be obtained by using the information
in Figure 2.2 and by making some assumptions concerning the
relative level of excitation of the modes. For this example,
the quite plausible assumption will be made that the observed
amplitude of an oscillation is independent of ℓ and m. It
will further be assumed that the amplitudes are inversely
proportional to the cube of the frequency in order to reproduce the essentially flat power spectrum that is observed.
The appropriate information to be derived from Figure 2.2 is
the number of modes per given frequency interval. This information and these assumptions regarding excitation were used
to generate a theoretical power spectrum to compare with the
observations of Brown, Stebbins and Hill.(42,52) Both the
resulting theoretical power spectrum and the observed power
spectrum are shown in Figure 2.3. The agreement of the theoretical and observational spectra in Figure 2.3 is quite good,
and thus from these observations alone, it is not possible to
conclude at this time that only low order modes are excited.

There is a second feature of the theoretical power spectrum in Figure 2.3 which should be noted. In addition to
reproduction of the structural form of the peaks, the theoretical treatment also reconstructs the continuum or background.
This suggests that in the case of this interpretation, which
does not impose any restriction on ℓ and m in the excitation
process, the power ascribed to noise by Brown, Stebbins and
Hill (42,52) and discussed in §2.2.1 is, to a large measure,
solar in origin and associated with oscillations. This
conclusion is also supported by the findings of Hill and
Caudell (43).

A distinction between the theoretical spectra obtained
for various assumptions regarding ℓ is not possible at this

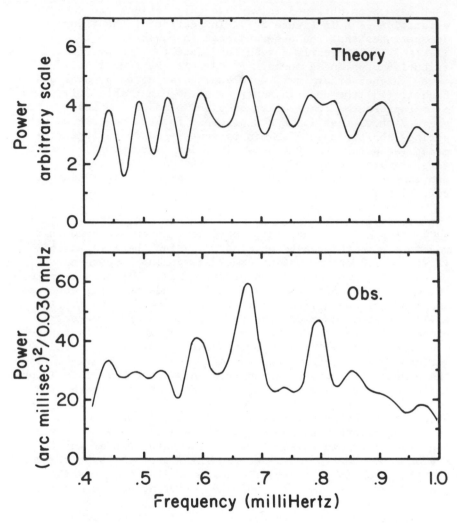

Fig. 2.3. Observed and predicted power spectra of solar oscillations for the diameter type observations. The
observed spectrum is from Brown, Stebbins and Hill
(42,52) . The predicted power spectrum was generated
by assuming that amplitudes of the oscillations
were independent of ℓ and m but decreased as the
fourth power of the frequency, the latter assumption being made in order to reproduce the observed
essentially flat power spectrum. The significant
feature to note is the quite good correspondence
between the number and locations of the peaks in
the two spectra.

time. However, the observed spectrum is not inconsistent with
that which might be expected. Thus, the peaks in the observed
power spectrum may be interpreted in terms of global modes of
solar oscillation.

2.3. Oscillations with Periods Longer than One Hour

Oscillations with periods near 2 h 40 min have been detected using velocity observations by Severny, Kotov and Tsap
(47), Kotov, Severny and Tsap (61) and Brookes, Isaak and
van der Raay (46). Velocity amplitudes of a few meters per
second were   reported and those oscillations observed August-
September 1974 at the Crimean Astrophysical Observatory were
found to be nearly in phase with the observations of March
1975 (47). The detection of these oscillations has continued
and Kotov, Severny and Tsap (61) reported that the oscillations observed in 1974, 1975 and 1976 were also found to be in
phase. Further, they reported the detection of these oscillations both in brightness changes and in measurements of velocity amplitude.

It should be noted here that Dittmer, Scherrer and Wilcox
(49) have recently reported negative results in their attempt
to detect these oscillations using velocity measurement techniques similar to those of Severny, Kotov and Tsap (47). These
contradictory results have not yet been resolved.

The most compelling evidence for the global nature of
this type of oscillation is the observed degree of coherency.
It has been noted in several studies, though, that the close-
ness of 2 h 40 min to an integral fraction of a day (1/9) is
such that the possibility that systematic diurnal effects are
operative must be considered. Worden and Simon (63) have
also examined the possibility that these oscillations represent the passing of large-scale solar velocity cells (supergranulation) through the field of view of the observing
instrument as the sun rotates. They concluded that although
this mechanism does not produce apparent oscillations that
remain in phase over long intervals of time, it could produce
the main features of the observed power spectrum for periods
near 2 h 40 min.

Although the work of Worden and Simon appears to offer a
reasonable mechanism for the origin of the 2 h 40 min oscillation, certain questions may be raised about their proposition.
Brookes, Isaak and van der Raay (46) have noted that the velocity amplitude produced by such a model is not large enough
to account for the observations of the 2 h 40 min oscillations. In addition, their interpretation raises a rather
puzzling question. If the apparent long term stability, i.e.

phase coherence, of the oscillation is nonsolar in origin,
then must not the oscillation itself be nonsolar in origin?
However, we would then be faced with finding a mechanism that
could explain how oscillations of nonsolar origin can have
properties nearly identical to those expected from a solar
phenomenon: namely, the rotation of supergranulation cells.
It is apparent that insight into the origin of the 2 h 40 min
oscillation must await further study, both observational and
theoretical.

An even longer periodic phenomenon has been found in the
Princeton solar oblateness data of Dicke and Goldenberg (63).
These data have been reanalyzed in an effort to understand the
sizable day-to-day variation which was discovered by Dicke (64,
65). The results of the analysis demonstrate a periodic variation in the apparent oblateness of the solar image having an
amplitude of about 2×10^{-5} of the solar diameter and a synodic period of 12.64 ± 0.12 days. The understanding of this
periodic structure must also wait for further observations and
theoretical analysis. However, a possible interpretation of
considerable interest proposes that this periodicity is due to
the beating of oscillatory modes that would have identical
frequencies except for the quite small differences produced by
the solar rotation (66,67). This interpretation is discussed
in more detail in §7.2.5.

3. Intercomparison of Observational Evidence on Solar Oscillations

A challenging problem in the study of solar oscillations
has arisen in the intercomparison of various observations. If
all studies of possible solar oscillations utilized identical
methods, disagreement in the interpretation of observational
results could be attributed to systematic errors generated,
for example, by the earth's atmosphere or the telescope or to
insufficient care in the data reduction. However, when the
studies implement a variety of instrumental parameters, such
as differing spectral lines, or when they examine different
fractions of the solar disk through the use of dissimilar
observational techniques, the intercomparison of the observational results may become rather complex. In these cases,
the sensitivity of each observational technique to solar oscillations and the influence of the instrumental parameters upon
this sensitivity must be determined. Insufficient consideration of these two general questions has led to several interesting paradoxes. These paradoxes have been encountered where
predictions based upon diameter measurements were not confirmed
by observations of intensity and velocity and predictions from
velocity measurements were not confirmed by intensity measurements. The following sections review the different

observational techniques used to date, present several paradoxes that have appeared as a result of using currently
accepted theory, and outline a procedure for intercomparison
of the observational results which will illuminate the weak-
nesses in the theory.

3.1. Different Types of Measurements

The observational techniques utilized in the study of
possible solar oscillations essentially involve the measurement of three classes of phenomena: (1) velocity of the solar
surface in the direction of the observer; (2) intensity
changes in the continuum and the absorption lines; and (3)
motions of the apparent limb, i.e. edge, of the solar disk.
Each of these categories provides a very large number of
opportunities for variation.

3.1.1. Velocity Observations The measurement of line of
sight velocity has been used with considerable success in the
study of solar rotation and in such studies as Leighton's
discovery of the five minute oscillation (23). This type of
measurement searches for Doppler shifts in spectral lines
using a grating spectrograph or resonance scattering techniques. Brookes, Isaak and van der Raay (46) and Grec and
Fossat (45) have employed resonance scattering in their work.
These measurements have typically been performed by examining
the entire disk at one time due to the low level of signals
produced by resonance scattering. Although statistics may be
improved by this method, the utility of these measurements
becomes considerably restricted, as will be discussed later.
The grating spectrograph has been used in work by Severny,
Kotov and Tsap (47); Dittmer, Scherrer and Wilcox (49);
Deubner (34); and Rhodes, Ulrich and Simon (36). The measurements in these cases are not confined to the concurrent
observation of the entire disk; a variety of spatial geometries have been used in examining the properties of the
oscillations. The most dramatic is that of the k-ω plot for
the five minute oscillation (see §2.1).

In addition to the spatial geometry of the detector, two
other variables to be considered are the specific spectral
line chosen for the Doppler measurements and, since the spectral lines all have finite widths, the location within the
broadened line. The information obtained through these measurements reflects the properties of the solar atmosphere at
a particular elevation. Thus the interpretation of such velocity observations requires knowledge of this effective height
and of the influence of oscillations upon the atmosphere at
this elevation.

3.1.2. Intensity or Temperature Observations Observations of intensity changes in the continuum and in spectral lines have often been used in the search for solar oscillations. Examples of this type are found in studies by Livingston, Milkey, and Slaughter (48); Musman and Nye (50); Beckers and Ayres (51); and Worden and Simon (62). These measurements all examine the sun directly. Indirect measurements were made by Deubner (68) in a study of planetary albedos.

Observations of this type examined the solar disk with a variety of spatial configurations, ranging from the whole disk measurements of Deubner (68) to the multigrid geometry of Musman and Nye. (50) Thus, spatial filtering is an additional factor which, as in the case of the velocity measurements, must be considered in detail. Intensity changes observed at various parts of different spectral lines, as well as in the continuum, reflect the properties of oscillations at a certain elevation in the solar atmosphere in much the same manner as do velocity observations. Thus the interpretation of such studies requires an understanding of spatial filtering as well as knowledge of the effective height and the influence of oscillations in the atmosphere.

3.1.3. Diameter Observations: Motions of the Solar Limb In recent years, a third type of technique has been introduced to the study of solar oscillations. This technique involves the measurement of the separation of two diametrically opposite limbs of the solar disk as a function of time. For these measurements a telescope is used which has been specifically designed and built for making precise distance measurements between stellar images in the daytime for testing theories of gravitation (see (69) for a discussion of the telescope's properties). This telescope produces an image of the sun with a minimum degradation. The separation of the two edges of the sun are measured interferometrically. Oscillations observed with this equipment have been referred to as oscillations in the apparent solar diameter (42). The use of the term "apparent" is necessary because, in any operational definition of a point on the solar limb, the actual intensity curve at the limb has a strong influence on the edge which is subsequently defined. Thus a change in the solar diameter could be a combination of an actual change and an apparent change produced by a modification of the intensity or limb darkening function, such as that produced, for example, by sunspots or the granulation on the sun's surface caused by convection.

A technique was pioneered at SCLERA for the study of solar oblateness utilizing a means of precise definition of the position of the solar limb. This method is described in

detail by Hill, Stebbins, and Oleson (70) and may be thought
of as comparing the intensity observed through an aperture
completely on the disk with the intensity observed through a
slot overlooking the edge of the disk. The aperture intensity is not very sensitive to changes in aperture position on
the disk, but the slot intensity is essentially proportional
to how much of it overlaps the solar disk. The edge in this
case would be defined as the position of the slot which
yielded an intensity equal to the aperture intensity. Formal-
ly, this technique may be conceptualized as a process of multi-
plication of the observed solar limb darkening function by an
appropriately chosen weighting function with subsequent integration of the result over some specified range near the sun's
limb. The range of integration is then moved back and forth
radially until a position is found at which the integral
vanishes. This position is defined as the edge or limb of the
solar disk. Thus, at the solar edge we can write

$$\int_{-a}^{a} G(r) \, W(r) \, dr = 0, \tag{3.1}$$

where G is the solar limb darkening function, W is the weighting function and a is half of the range of integration. The
parameter a is also referred to as the scan amplitude since it
specifies how much of the limb is examined or scanned in the
integration. In practice, the weighting function is a sinusoidal function given by

$$W(r) = \cos\left[2 \arcsin\left(\frac{r-r_0}{a}\right)\right], \tag{3.2}$$

where r_0 is the center of the range of integration. With this
weighting function, the integral transform used in the definition, i.e., equation (3.1), may be recognized as a finite
Fourier transform of the limb darkening function. Thus it is
referred to as the finite Fourier transform definition of an
edge on the solar disk, or simply the FFTD.

Two properties of the FFTD are important in the current
context. The first is that the edge position is dependent
upon the brightness distribution at the extreme solar limb,
and not explicitly upon the position or motion of mass elements of the sun. Thus the manifestation of solar oscillations
at the solar limb must be understood in order to interpret such
observations. Equations (3.1) and (3.2) also indicate that,
just as with the other observational techniques, spatial
filtering is an integral part of the definition. The characteristic radial distance is given by scan amplitude a while
the characteristic distance perpendicular to the radius is
specified by the fraction of the circumference used in the
measurement of G(r).

3.2. Paradoxes Encountered on Intercomparison of Differing Observations

The intercomparison between results of the three broad categories of observational methods theoretically capable of detecting solar oscillations as described in §3.1 has, in the traditional scientific spirit, been approached from the simplest consideration of the system. Serious discrepancies have developed in the comparison of limb motion vs. velocity, limb motion vs. intensity fluctuation, and intensity fluctuation vs. velocity. These are reviewed in the following sections in order to convey a feeling for the magnitude of these paradoxes and to show that their source is not traceable to a single set of observations.

3.2.1. Fluctuations in Limb Position vs. Velocity The first intercomparison (71) was made between the limb motion observations of Brown, Stebbins and Hill (42) and the whole disk velocity observations of Brookes, Isaak and van der Raay (46). The simplifying assumptions made in this intercomparison are: (1) the effects of spatial filtering can be neglected and (2) the observed limb shifts in the SCLERA observations are in fact real translations of the limb. The first assumption cannot be justified since the limb shift measurement examines an approximately 200 x 100 (arcsec)^2 segment, equal to about 0.2% of the solar surface, whereas these velocity measurements involve the entire disk. The discussion in §3.1.3 with regard to the sensitivity of the FFTD to changes in the limb darkening function demonstrates the possibility that the second assumption may also be invalid.

Recognizing these limitations, it is nonetheless of value to review the intercomparison. Brookes, Isaak and van der Raay (46) reported a line-of-sight velocity of 80 cm/sec at a period of 58 minutes, corresponding to a diameter amplitude of 1.4 milli-arcsec for this oscillation. The measurements by Brown, Stebbins and Hill (42) give at this period a diameter amplitude of 6 milli-arcsec, a discrepancy of 4.3 in relative amplitudes.

This type of comparison can also be made, under similar assumptions, using the whole disk velocity measurements of Grec and Fossat (45) and the SCLERA limb shift measurements. In this case the oscillation period of 28 min is chosen for the comparison. Grec and Fossat do not observe an oscillation and report an upper limit in the velocity of $2m^2/sec^2/$ mHz. Converting this power to the same frequency interval used by Brown, Stebbins and Hill (42), 0.0305 mHz, the resulting upper limit in velocity corresponds to a diameter amplitude of less than 0.3 milli-arcsec. The measurements of

Brown, Stebbins and Hill (42) report a diameter amplitude at the period of 28 min of 5.6 milli-arcsec, which is larger by a factor of 19 than that of Grec and Fossat.

Other comparisons which lead to similar discrepancies are possible, demonstrating the paradox resulting from the correlation of velocity and limb shift measurements.

3.2.2. Apparent Limb Motion vs. Intensity Fluctuations

The intercomparison of limb motions as observed at SCLERA and direct intensity fluctuations is more complex than the velocity problem previously discussed. The comparison of these two types of measurements necessitates the formulation of some assumptions regarding spatial filtering and a model for the effects of solar oscillations upon the location of the apparent limb and upon intensity changes in the continuum and in spectral lines. This demands not only a theoretical treatment of wave propagation for a determination of an oscillating system's thermodynamic properties, but also the more difficult theoretical consideration of the manner in which spectral lines are affected by these properties. It should not go unnoticed that the observations which are being utilized in the search for solar oscillations provide the only test for this theoretical analysis of spectral lines.

An intercomparison of limb motions and intensity fluctuations may be derived from the work of Livingston, Milkey and Slaughter (48) and Hill, Caudell and Rosenwald (71). The latter authors present the results of a general treatment of acoustic wave propagation in the photosphere and note that it is possible to resolve the apparent limb motion vs. velocity paradox within the framework of this general treatment. This leads, for example, to a temperature amplitude of 0.7 K at an optical depth $\tau = 0.1$ for an oscillation period of 58 min. Livingston, Milkey and Slaughter made whole disk measurements of intensity fluctuations in the spectral line CI 5380, and found no evidence for oscillations. They conclude that, in the event oscillations do exist, the temperature perturbation at the base of the photosphere, neglecting the effects of spatial filtering, must be less than 0.3 K at a period of 58 min. Although this difference is not as large as that encountered in the comparison of apparent limb motion vs. velocity, it does represent a discrepancy.

A similar type of intercomparison has been made by Beckers and Ayres (51). In their search for intensity oscillations they examined the central parts of the Ca K spectral line (3933A) for whole disk observations and for a small circular area at the center of the solar disk with a diameter of 5 arc min. For the whole disk and disk center measurements, they imposed an upper limitation of 0.02% and 0.05%, respectively, on the amplitude of intensity fluctuations having periods ranging from 30 to 80 minutes for the K_1 part of the Ca II spectrum (±0.3A off line center). Basing their work on that of Hill, Rosenwald and Caudell (72), Beckers and Ayres conclude that the apparent limb shift measurements of Brown, Stebbins and Hill (42) predict amplitudes of 0.5% in the intensity. This constitutes a discrepancy of a factor of 25 for the whole disk intensity measurements and a factor of 10 for those at disk center.

The work of Musman and Nye (50) offers a third comparison of this type. They examined the continuum radiation for fluctuations in intensity, but in this case the intensity was measured at 32 different positions on the solar disk, arranged in a 4 x 8 grid at disk center. The distance between nearest neighbors was 125 arcsec. The spatial filtering in these observations was quite complex due to the manner of sampling of intensity at the 32 positions and the way these intensity measurements were normalized after each set of 32 measurements. However, this complexity was not considered, and Musman and Nye conclude that amplitude per frequency interval was less than 2×10^{-4} in intensity or 0.3 K in brightness temperature.

The upper limit of 0.3 K is similar to that obtained by Livingston, Milkey and Slaughter (48) using the CI 5580 line and is again less than the 0.7 K amplitude implied by the limb shift observations.

3.2.3. Intensity vs. Velocity Fluctuations. Simultaneous observations of intensity and velocity for the five-minute oscillation offer a possible method of testing the theory used in the intercomparisons of the previous two subsections. This theory correlates acoustic wave propagation in the solar atmosphere and the manifestation of the oscillations in spectral lines. Although this involves areas of physics relatively well understood, serious discrepancies have been discovered.

The simultaneous intensity and velocity measurements of the five-minute oscillations of Leighton, Noyes and Simon (73) and Tanenbaum et al. (74) were intercompared by Hill, Caudell and Rosenwald (75) on the basis of the acoustic wave treatment of Hill, Rosenwald and Caudell (72). It was found that the ratio of intensity change to velocity amplitude was an order of magnitude below that predicted by line formation theory for the base of the photosphere and for the chromosphere.

It is particularly significant that this discrepancy was found for the same chromospheric Ca K line used by Beckers and Ayres (51). The fact that these theoretical analyses are insufficient in the comparison of intensity and velocity fluctuations for the five-minute oscillation indicates that caution should be exercised in the formation of conclusions regarding the apparent limb shift measurements.

A similar discrepancy in simultaneous intensity and velocity measurements of the five-minute oscillation has been discovered in the CI 5380 line. Hill, Livingston and Caudell (76) found the temperature sensitivity of this line to be a factor of 35 below that predicted by Livingston, Milkey and Slaughter (48), again indicating the uncertainty of the applicability of the latter researchers' results to the work of Hill, Caudell and Rosenwald (71).

A third intercomparison between the fluctuations in the continuum and the velocity of the five-minute oscillations has been made by Hill, Caudell and Rosenwald (75). They analyzed the observations of Tanenbaum et al. (74) and, via the acoustic wave analysis of Hill, Rosenwald and Caudell (72), found that the temperature sensitivity of the continuum radiation is a factor of 35 below that for a system radiating as a black body. However, it was the temperature sensitivity of the black body system that was used by Musman and Nye (50) in obtaining a temperature amplitude of 0.3 K from a 2×10^{-4} change in intensity. This discrepancy again suggests that there are problems inherent in the practice of comparing apparent limb motion with intensity fluctuations.

3.3. A "Recipe" for Intercomparison: An Approach to the Resolution of the Paradoxes

It is apparent from the discussion in §3.2 that the intercomparison of velocity, intensity fluctuations, and apparent limb shifts is wrought with many uncertainties. There are essentially three types of problems which may be decoupled and analyzed independently. These are: (1) the effects of spatial filtering, (2) the general treatment of acoustic wave propagation in the solar atmosphere, and (3) the general problem of spectral line response to oscillations.

The effects of spatial filtering on the sensitivity of each observation must be carefully analyzed before the results may be interpreted. Spatial filtering analysis prior to the observation would probably result in a better designed and directed program. Sensitivity can be studied either by numerical calculations or by analytical techniques, although the analytical techniques somewhat facilitate the examination

of more general sensitivity properties. Examples of this work
are discussed in §4.

A systematic approach to the intercomparison of observations additionally necessitates a study of the role of the
sun's outer layer in the detection of oscillations. For such
a study, the eigenfunctions or, more specifically, the general
solutions of the wave equations in the photosphere, are necessary in order to specify pressure, density, temperature, and
velocity amplitudes as functions of position in the solar
envelope. An analysis of this type has been carried out by
Hill, Rosenwald and Caudell (72) and is discussed in §5.

A final step in systematic intercomparison concerns the
boundary conditions used to determine integration constants
in the general solution of the wave equations. Although boundary conditions have frequently been considered in stellar
pulsation theory, these previous treatments have not been
quantitatively tested by direct observation as has been the
case with research on solar oscillations. The first results
in solar seismology have dealt with the study of these boundary conditions as discussed in §6.1. Further quantitative
work of this type is required in order to specify, to the
requisite accuracy, the constants in the general solution of
the wave equations.

For those observations using spectral lines, it is necessary to examine the manner in which the oscillations are
reflected in the time-dependent shape of the spectral line.
The complexity of this interaction promises to make its study
difficult; however, §6.2 describes several recent advances in
this field, and it is likely that more will be forthcoming.
An adequate understanding of this system will no doubt come
only through a careful correlation of observations with theory.

4. Detector Sensitivity
and the Spatial Properties of Waves

The amplitude of an oscillation in the sun will not only
exhibit temporal periodicity but will also change in a periodic fashion from one point on the surface to another. The
"horizontal" scale or wavelength associated with this periodicity may range from 10^{-3} solar diameters upward until the
limiting case is reached where the amplitude is independent of
position on the surface, i.e., a spherically symmetric wave.

An observation performed with a given detector configuration will generally be most sensitive at a particular horizontal wavelength and possess a reduced sensitivity at other wavelengths. Thus the detector may be thought of as a spatial
filter in that only oscillations with certain horizontal wave-

lengths are preferentially observed. In practice, these effects of spatial filtering have been overlooked in the interpretation of observations on solar oscillations; this casts
considerable doubt upon conclusions reached in this manner.

In the following subsection, a formalism is presented
which has been developed for the evaluation of spatial filtering for several detector geometries which have been used in
the study of solar oscillations and the results of this analysis are presented in order to demonstrate the importance of
spatial filtering.

4.1. The Calculation of Spatial Filter Functions

The spatial properties of an oscillation as a function of
both height in the atmosphere and position on the sun's surface are given by the eigenfunction for that oscillation. The
position-dependent portion of the eigenfunction is given by
the spherical harmonic $Y_\ell^m(\theta,\phi)$, where ℓ and m are the principal and azimuthal order numbers of the spherical harmonic,
respectively, and θ and ϕ are the spherical coordinates with
the $\theta = 0$ direction aligned to the rotational axis of the sun.

The problem of calculating the spatial filter function in
this representation of a solar oscillation is thus reduced to
the integration of a particular Y_ℓ^m over the solar disk with an
appropriate weighting function. The weighting function would
reflect such factors as the size, shape, and location of the
detector aperture and the type of observation such as line-of-
sight velocity or line intensity. This integration can be
considerably simplified in many instances by taking advantage
of the properties of the Y_ℓ^m under a finite rotation of the
coordinate system. The subsequent integration over θ and ϕ
can be changed in a variety of cases to Fourier analyses of
the Y_ℓ^m and the geometry of the aperture. In this manner the
calculation of the spatial filter function for a particular
observation is reduced to the performance of a Fourier transform of the appropriate detector geometry and summing terms.
The evaluation of a Fourier transform is often a much simpler
operation than integration of Y_ℓ^m and weighting function in
each new observational set up.

Knowledge of the properties of the Y_ℓ^m under rotation of
the coordinates and the Fourier series of Y_ℓ^m is required in
order to implement the calculation. The former is to be found
in many textbooks on rotation operators (see for example
(77)). The Fourier analysis of the Y_ℓ^m has been obtained and
the new results are presented in Appendix A. This appendix
also discusses the use of the rotation operator and presents
the appropriate matrix elements for completeness.

Table 4.1

Velocity Sensitivities of Whole Disk Observations*

ℓ^m	0	1	3	5	7	9	11	13
0	6.67 E-1							
1		4.33 E-1						
2	2.98 E-1							
3		6.75 E-2	8.72 E-2					
5		5.36 E-3	9.03 E-3	1.21 E-2				
7		2.50 E-3	2.60 E-3	2.87 E-3	3.92 E-3			
9		1.04 E-3	1.07 E-3	1.12 E-3	1.26 E-3	1.72 E-3		
11		5.20 E-4	5.28 E-4	5.47 E-4	5.32 E-4	6.50 E-4	9.05 E-4	
13		2.92 E-4	2.96 E-4	3.03 E-4	3.15 E-4	3.38 E-4	3.82 E-4	5.30 E-4

*The sensitivity is zero if no entry is made in the table and for those values of ℓ and m that are $\ell \leqslant 13$ and not shown.

4.2 Examples of Spatial Filter Functions

Two types of observations which have been made at different observatories are those which involve whole disk and differential velocity measurements. In the whole disk measurements, the light from the entire solar image is passed through a velocity spectrometer. Brookes, Isaak and van der Raay (46) and Grec and Fossat (45) have utilized this method. In the differential velocity observations, such as those of Dittmer, Scherrer and Wilcox (49) and Kotov, Severny and Tsap (61), the velocity observed in a central circular area of the disk is subtracted from the velocity observed in an outer annulus. It is of interest to examine these filter functions in some detail.

4.2.1. Whole Disk Velocity Measurements

The general expression for the whole disk velocity spatial filter function is given by equation (A10) of Appendix A with $\eta = 3$ and $q = 0$. This has been evaluated where the $_4F_3$ function reduces to a $_3F_2$ function. The results are listed in Table 4.1 as a function of ℓ and m. Inspection of the results in this table indicates that this technique has a high sensitivity for only $\ell = 0$, 1 and 2 and that the sensitivity drops off as ℓ^{-3}. For whole disk measurements in radiation intensity, the spatial filter function including the effects of limb darkening is, to a good approximation, given by equation (A10) with $\eta = 3/2$ and $q = 0$. In this case, the sensitivity is again high only for $\ell = 0$, 1 and 2 and drops off approximately as $\ell^{-3/2}$.

4.2.2. Differential Velocity Measurements

The computation of the spatial filter function for differential velocity measurements is obtained by combining the Fourier analyses of the associated Legendre polynomial (equations A10 and A11), the Fourier analysis of the weight or window function, and the rotation matrix elements (equation A9). These expressions have been evaluated for the geometry used by Dittmer, Scherrer and Wilcox (49), where the radius of the circular aperture and the inner and outer radii of the annulus are respectively 0.5 $R_\odot$, 0.55 $R_\odot$ and 0.80 $R_\odot$ (78), and $R_\odot$ is the radius of the solar disk. The results of this analysis are plotted in Figure 4.1 as the root mean square sensitivity, where the average has been taken over m. The sensitivities for $m < \ell$ increase monotonically from $m = 0$ to $m = \ell$, with the $m = 0$ value about 1/2 of the $m = \ell$ value. Similar results are expected for the observations of Kotov, Severny and Tsap (61).

Examination of Figure 4.1 shows that this type of observation is useful for $\ell \lesssim 12$, where the average sensitivity is about 0.15. For values of ℓ greater than 12, the sensitivity drops off approximately as ℓ^{-2}.

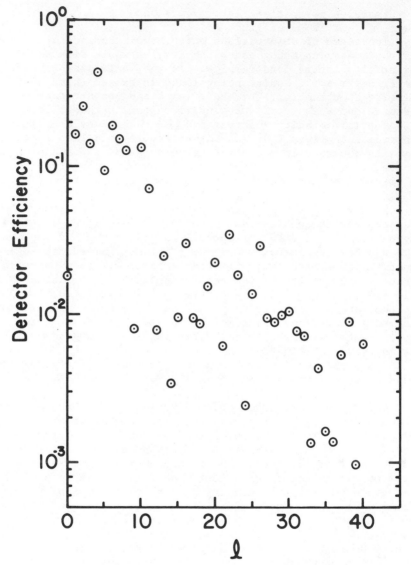

Fig. 4.1. The spatial filter function for the differential
velocity measurements of Dittmer, Scherrer and
Wilcox (49) when observing p modes. The velocity
sensitivity plotted is the root mean square velocity averaged on the azimuthal order number, m. The
salient feature to note is that the sensitivity
fluctuates around a mean value of 0.15 for $\ell \lesssim 12$
and decreases as ℓ^{-3} for values above 12. The
spatial filter function for the differential velocity measurements of Kotov, Severny and Tsap (61)
should exhibit similar properties for p modes.

<u>4.2.3. Limb Motion Observations</u> The spatial filter functions for the FFTD are expected to be dependent upon ℓ, m, and position angle of the aperture on the solar disk, and certain other properties of the eigenfunction, such as those that reflect the boundary conditions. It is also expected to be different for p and g mode oscillations. In order to display the general character of the spatial filtering encountered in the FFTD, it will suffice to examine one of the simpler cases of equatorial diameter measurements at an oscillation period of 25 min. Fortunately, most of the relevant SCLERA observations have been made at or near the equator.

Restriction of the measurements to the equatorial region results in a decoupling of the θ and ϕ integrations in the filter function calculations. Thus, there is no need to introduce the complexity of the finite rotation operator with its matrix elements containing Jacobi polynomials (see discussion in Appendix A). Also at a period of 25 min, the eigenfrequency spectrum for a standard solar model consists of only p modes that are low order with respect to ℓ, as shown in Figure 2.1. This indicates that at this particular period, the spatial filter functions obtained by using the solutions of Hill, Rosenwald and Caudell (<u>72</u>) should be quite relevant to the corresponding observations.

The other essential parameter that must be specified in these calculations is that of the length of the slit parallel to the edge of the solar disk. In the observations of Brown, Stebbins, and Hill (<u>42,52</u>) and Hill and Stebbins (<u>54</u>), this length was 100 arcsec, or approximately one tenth of $R_\odot$.

The calculation of the spatial filter function was accomplished in several steps. First, the integration over θ was achieved by combining the results of the Fourier analysis of the window function in θ with the Fourier analysis of the Y_ℓ^m (Equations A10 and A11). These results were subsequently combined with the FFTD analysis of the $\delta I e^{im\phi}$, where the δI was given by Hill, Rosenwald and Caudell (<u>72</u>). The final results of these analyses are shown in Figure 4.2 as a function of ℓ for the β_+ solution and two different scan amplitudes; a root mean square average has been made over m.

It is important to note that the sensitivity of the SCLERA type measurements for a sizable range in ℓ is significantly larger than that given for $\ell = 0$ by Hill, Rosenwald and Caudell (<u>72</u>). Taking this into account, the approximate values of ℓ for which the sensitivity has decreased by $2^{-\frac{1}{2}}$ for the 27.2 and 6.8 arcsec scan amplitudes are 30 and 60, respectively. By interpolation, a corresponding half power

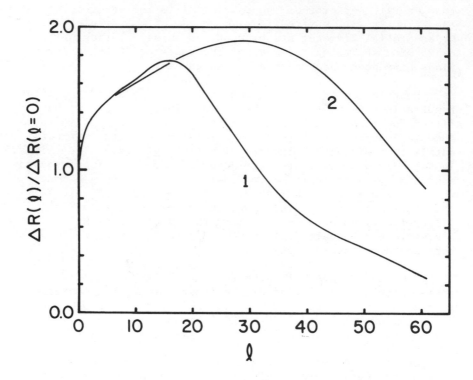

Fig. 4.2. Spatial filter functions for equatorial diameter
type measurements at an oscillatory period of 25
min. Curves 1 and 2 were computed for 27.2 and 6.8
arcsec scan amplitudes, respectively. The $\Delta R (\ell)$ is
the apparent change in the location of the solar
limb due to intensity changes produced by the β_+
component of an eigenfunction describing a solar
oscillation, of principle order ℓ, as observed by
the FFTD. A root mean square average has been performed over m. The length of the slit parallel to
the solar limb is 100 arcsec.

point of approximately 45 is obtained for the observations of
Brown, Stebbins and Hill (42,52) also discussed in §6.5.

5. The Influence of the Outer Layers of the Sun on the Detection of Solar Oscillations

A complete theoretical solution for the pulsation of a
star generally requires the use of a wave equation describing
wave propagation within the star and the specification of the
star's boundary conditions, both at the center and in the
envelope. Although boundary conditions at the outer portions
of the star have received considerable attention (see, for
example, 25, 28, 32, 79, 80, 81), a number of uncertainties
still remain in this area. This is attributable in large measure to the secondary role that the boundary conditions play
in determining the eigenfrequencies of pulsation and in the
stability of a star against pulsation. These two observables,
eigenfrequencies and stability, show little relationship to
outer boundary conditions, diminishing their application in
testing hypotheses dealing with the boundary.

On the other hand, most of the observations that are
currently available on solar oscillations deal with physical
quantities measured at heights in the atmosphere where the
boundary conditions have traditionally been set. Consequently
the anticipated properties may be in considerable error,
leading to misinterpretation of the observations. This is
particularly true of the diameter measurements at SCLERA.

Compared to eigenfrequencies and solar stability against
pulsation, study of the role of the sun's outer layers in the
detection of oscillations has been neglected. Studies of
solar stability and eigenfrequency calculations have not been
encumbered by the lack of a good understanding of the relevant
outer boundary conditions. This is not the case, however,
when the investigation involves the eigenfunctions, the functions which describe the temporal and spatial properties of
the oscillations. At or near the boundary, their properties
are extremely sensitive to the applied surface conditions;
small errors can significantly modify the predicted properties
of observable quantities.

It has often been noted that the eigenfunctions obtained
with commonly used boundary conditions poorly describe the
propagation of photospheric waves (14,32,82). The difficulties
arise primarily from two sources; the extreme complexity
presented by the presence of the chromosphere and corona, and
until now, the lack of a means to observationally test a
particular model.

5.1. The Second Solution of the Wave Equation and Its Enhancement Factor

It has been noted by Hill, Caudell and Rosenwald (71) that a change of a few parts per thousand in the limb darkening function can account for the SCLERA results whereas changes of parts per million in the diameter can produce the line-of-sight velocity amplitudes. Further, the assumption of the correctness of both types of measurements implies the existence of a strong enhancement factor for certain observable quantities which, if real, has not previously been accounted for in theoretical considerations.

Hill, Caudell and Rosenwald (71) have demonstrated (see (72) for a more complete treatment) that one of the two solutions of the wave equation does possess a considerable enhancement factor as an intrinsic property. In the following discussion the two solutions will be noted by β_+ and β_-; the β_+ solution will exhibit the enhancement factor. As pointed out by Ulrich (32); Zhugshda (81); and Stein and Leibacher (82), both solutions will exist in an atmospheric layer of finite thickness. However, the various treatments of the photosphere, chromosphere, and corona in theoretical studies of oscillations have in general led to the conclusion that in the photosphere the β_+ solution is negligible relative to the β_- solution. It is for this reason that the enhancement factor of the β_+ solution has not been apparent in previous theoretical considerations.

5.2. Specification of the Relative Amplitudes of the Two Solutions through Observation

The establishment of proper boundary conditions is an essential element and, as noted in §5.1, errors in these boundary conditions strongly influence the theoretical predictions in the photosphere. However, instead of attempting to choose exact boundary conditions and predicting the observations, it is possible to leave unspecified to within two constants of integration the mechanical and thermal boundary conditions and to numerically obtain specific solutions to the nonadiabatic wave equation. The resulting general solution is of the form

$$\xi = A_+ \, \xi_+ + A_- \, \xi_- \, , \tag{5.1}$$

where A_+ and A_- are the two constants of integration, $\xi = \delta r/r$, δr is the Lagrangian perturbation in the radius r, and ξ_+ and ξ_- are respectively the β_+ and β_- solutions of the wave equation. This can subsequently be applied to various observations. Through such analysis, constraints may be placed on the unspecified constants of integration. New information may be obtained in this manner about conditions near the surface, illustrating the value of seismic sounding as a valuable probe in the study of solar boundary conditions.

5.3. Numerical Solutions of the Wave Equation

Solutions of the wave equation may be found in the literature (e.g., 14,83). Unfortunately, these are not general solutions in that certain boundary conditions have been assumed in obtaining a particular solution. In accordance with the premises discussed in §5.2, Hill, Rosenwald and Caudell (72) generated several sets of general solutions where two constants of integration remained unspecified. These solutions were obtained by detailed numerical calculations of linearized conservation equations incorporating a model of the solar atmosphere. The nonadiabatic term of the conservation of energy equation was treated in the Eddington approximation, generalized to the three-dimensional case after Unno and Spiegel (84).

It is important to note that the treatment of the nonadiabatic term makes the "κ-mechanism" (see (5)) for driving oscillations (85) an integral part of these numerical calculations and contributes prominently to the properties of the final solutions. This has subsequently become central to several considerations (see discussion in §6.2 and §7.1.1).

5.4. Various Observational Sources for Constraints on the Relative Amplitudes of the Two Solutions

Two constants of integration remain to be determined: the first may be thought of as a scale factor and the other as the ratio of the two amplitudes. The ratio, which is the most important quantity to establish, may be complex, leading to three unknowns. This necessitates the formulation and solution of three simultaneous equations. Fortunately, there are several combinations of observables which can be used to set up the three simultaneous equations. One set of observables involving the five-minute oscillation includes the magnitudes and relative phase of simultaneous intensity and velocity measurements within a spectral line or a combination of a spectral line and continuum radiation. A discussion of the use of this procedure is provided in §6.2. Another set of observables contains the magnitudes and relative phase of limb motion and velocity observations. Work in this direction has already met with some success as is described in §6.1. Thus, for the first time we have an opportunity to study the manifestation of the outer boundary conditions used in stellar pulsation theory.

6. Solar Seismology:
The First Findings

The emphasis of theoretical work on solar oscillations has addressed the model dependence of the eigenfrequencies (8, 55,56,59). A conclusion that might easily be drawn from such a concentration of effort is that studies in solar seismology are relevant only to the determination of the internal structure of the sun. However, such a conclusion is not justified in light of recent work which is discussed in some detail in the following sections. These studies have produced significant results relating to the sun's outer layers. Initial results concern the boundary conditions used in stellar pulsation theory, the temperature perturbations associated with oscillations, evidence for nonlinear effects in the oscillations, new constraints on the convection zone, and a characteristic horizontal scale for the five-minute oscillatory mode.

6.1.   Boundary Conditions Used in Stellar Pulsation Theory[†]

Studies undertaken in the last two years have suggested that testing of solar boundary conditions associated with oscillations may, at last, be possible. It was noted by Hill, Caudell and Rosenwald (71) that traditional treatments have offered incomplete interpretations of solar pulsation; they have described inconsistencies resulting from these interpretations (see §5) in relation to recent observations discussed in §3. The observations which most clearly point to inadequacies in the traditional treatment have all been obtained at SCLERA; this fact has led some workers to suggest that these observations may be in error. However, the data reviewed in §2.2 do not support this hypothesis. The following discussion demonstrates the insufficiency of traditional models of solar pulsation in the interpretation of the SCLERA observations.

The inconsistency between the limb motion and velocity observations may be clarified to a greater extent than is indicated in §3.2.1. In that section it is stated that discrepancy factors of 4.3 and 19 were obtained upon comparison of whole disk velocity measurements with limb motion measurements. The detailed analysis of the respective spatial filter functions (§4.2.1 and §4.2.3) shows that these conclusions are valid only if oscillations with the lowest order values of ℓ are excited; the sensitivity of the whole disk measurements is

---

[†]Contributed by: H.A. Hill, Dept. of Physics, U. of Arizona
                  R.D. Rosenwald, Steward Obs., U. of Arizona
                  T.P. Caudell, Dept. of Physics, U. of Arizona

proportional to ℓ^{-3}, whereas the sensitivity of the Brown, Stebbins and Hill (42,52) observations remains high for values of $\ell \lesssim 45$. However, the analysis of the 1973 solar oblateness observations (43) shows that modes with $\ell \approx 45$ are excited at periods near the comparison period of 58 min. At the other comparison period of 28 min used in §3.2.1, the theoretical spectrum for the modes of oscillation derived from the standard solar model (see 2.2) has modes with

$$\ell \lesssim 12. \tag{6.1}$$

Thus, both discrepancy factors could be the result of the excitation of oscillatory modes with $\ell > 0$ and different detector sensitivities for these modes.

An intercomparison of the limb motion observations and differential velocity measurements also results in a discrepancy. However, it is much more difficult to attribute this to a simple difference in spatial filter functions. The spatial filter function for the differential velocity observations of Dittmer, Scherrer and Wilcox (49) and Dittmer (78) has been evaluated and plotted in Figure 4.1. Examination of this figure indicates that sensitivity remains relatively high for $\ell \lesssim 12$ with an average value of about 0.15. In addition, it is only weakly dependent upon m. Thus, if the standard solar model yields a frequency spectrum typical of the sun, then the values of permissible ℓ are restricted to the same range, $\ell \lesssim 12$, as previously noted. Since the sensitivity of the limb motion observations by Brown, Stebbins and Hill (42,52) remains high for $\ell \lesssim 45$, a discrepancy arising from the intercomparison of these two sets of observations at a period of 28 min would not be due to spatial filtering effects.

The velocity reported by Dittmer, Scherrer and Wilcox at this period is less than 1.2 m^2/sec^2/mHz. Correcting for the average efficiency factor of 0.15 and converting to the frequency interval (0.0305 mHz) used by Brown, Stebbins and Hill (42,52), the resulting velocity corresponds to an oscillation amplitude of 0.48 milli-arcsec. The measurements of Brown, Stebbins and Hill (42,52) indicate a diameter amplitude of 5.5 milli-arcsec at this period. Clearly these two results are not compatible and exhibit a discrepancy of 5.8.

Another major difference is found in an analysis of the 1973 solar oblateness observations by Hill and Caudell (43). Traditional treatment of the boundary in pulsation theory asserts that the observed amplitude of an oscillation should not depend upon the scan amplitude in the FFTD (see §3.1.3). Results of the solar oblateness analysis, however,

contradicted this assertion in a frequency range in which
effects of changes in the spatial filter function should have
been insignificant, assuming the essential accuracy of the
description of the solar frequency spectrum offered in Figure
2.2.

The phase coherency of the solar oscillations in time,
as discussed in §2.2, clearly supports the classification of
these oscillations as global modes of the sun. Thus, the
previously discussed intercomparisons, which have led to discrepancies using the standard treatment of the outer boundary
conditions, may be used as noted in §5.2 for the specification of the amplitudes of the two solutions, A_+ and A_-, of the
general solution given by equation 5.1. The results of this
analysis are listed in Table 6.1.

The result of an analysis of data pertaining to the five-
minute oscillatory mode is also listed in the table; it was
accomplished by Hill, Caudell and Rosenwald (75) using the
simultaneous observations in continuum fluctuations and velocity of Tanenbaum et al. (74) and the appropriate β_+ and β_-
solutions discussed in §5. The result for oscillatory periods
near 33 min was obtained through the use of the dependence of
observed oscillatory power on scan amplitude found by Hill and
Caudell (43). Although this finding does not contribute to
the determination of (A_+/A_-), its significance lies in its relative immunity to inadequacies in either the theoretical treatment and/or the observational interpretation. The analysis of
the observations relevant to a 28 min period is nearly comparable in importance to that of the 33 min period in that the
probability of g mode contamination is quite small and the
results of the theoretical calculations are more reliable for
this analysis than for the five-minute mode.

The results in Table 6.1 are among the first to be obtained in solar seismology. The fact that the amplitude of
the β_+ solution is not negligible in the photosphere, i.e.,
A_+ is not much less than A_-, shows that the traditional treatment of the outer boundary conditions is not adequate. Even
more significantly, constraints have been placed on the relative amplitudes, A_+ and A_-, and consequently on the actual
boundary conditions on pressure, density, and temperatures
as "seen" by solar oscillations at the top of the photosphere.

The inadequacy of various theoretical treatments must lie
in part in the complex nature of the chromosphere and corona.
However, it may also be a manifestation of the manner in which
the oscillations are excited and, in turn, the manner in which
the chromosphere and corona are heated. For example, the
driving of the modes in the hydrogen ionization zone by the

Table 6.1

Magnitudes for Constants of Integration
Inferred from Observations*

Π min	$\|A_+\|$ $(\times 10^6)$	$\|A_-\|$ $(\times 10^6)$	$\|A_+/A_-\|$	Sources
58	$\simeq$ 0.47	$\simeq$ 1.0	$\simeq$ 0.5	a,b
	$\simeq$ 0.47	< 0.9	> 0.5	a,c
33	$\simeq$ 0.22			d,e
28	$\simeq$ 0.29	< 0.5	> 0.6	a,f
5	$\simeq$ 51	$\simeq$ 34	$\simeq$ 1.5	g,h

a. Brown, Stebbins and Hill (42)
b. Brookes, Isaak and van der Raay (46)
c. Grec and Fossat (45)
d. Hill and Stebbins (54)
e. Hill and Caudell (43)
f. Dittmer, Scherrer and Wilcox (49)
g. Tanenbaum et al. (74)
h. Hill, Caudell and Rosenwald (75)

*The constants of integration were evaluated for $\xi_\pm = 1$
at $\tau = 2.3 \times 10^{-3}$.

κ-mechanism favors by a considerable factor the β_+ solution over the β_- solution. This is because the driving is proportional to the amplitude of the density fluctuations, which are relatively larger for the β_+ solution. Furthermore, if these oscillations with a relatively large β_+ component are important in the heating of the chromosphere and corona, we then may be left with a highly coupled problem; the region which establishes the boundary conditions for the general solution is itself maintained in part by the β_+ solution.

6.2. Temperature Perturbations Associated with Oscillations[†]

The temperature associated with any phenomenon in a star must, like all of the thermodynamic variables, be inferred from the observable properties of the system. With the exception of the determination of the effective stellar surface temperature (i.e., the temperature of a black body equivalent in total luminosity), temperature estimation is a rather involved and convoluted process which has traditionally required detailed knowledge of spectral line formation in both emission and absorption.

Recent observations suggest that previous estimations of the amplitude of temperature perturbation associated with the five-minute oscillation may be in considerable error. This discovery, if correct, would have significant implications in that the model of the outer layers of the sun and hence, of stars in general, would be complicated considerably; current models assume that oscillations are unimportant in specifying the average properties of stellar atmospheres. It further appears that nonlinear stellar pulsation theory may be required in this area of solar physics. This may not only lead to a better understanding of the latter, but may also furnish a valuable testing ground for certain aspects of stellar pulsation theory.

Various methods of ascertaining the temperature perturbation associated with the five-minute oscillation have been utilized; some of these may be found in work by Leighton, Noyes and Simon (73), Holweger and Testerman (86) and Livingston, Milkey and Slaughter (48). Leighton, Noyes and Simon estimated temperature amplitudes by assuming that intensity changes in the absorption line were identical to those produced by a black body at the same temperature. Holweger and Testerman obtained estimates of temperature variation by utilizing the temperature sensitivity of the widths of

---

[†]Contributed by: H.A. Hill, Dept. of Physics, U. of Arizona
                    T.P. Caudell, Dept. of Physics, U. of Arizona
                    R.D. Rosenwald, Steward Obs., U. of Arizona

spectral lines. Livingston, Milkey and Slaughter inferred
temperature perturbations by evaluating the temperature sensitivity of the strength of an absorption line from line formation calculations. Heretofore, the reasonably close agreement
between these studies suggested that the physics of the temperature perturbation was fairly well understood.

However, the introduction of a new technique for estimation of temperature perturbations advanced the possibility
that the agreement between the aforementioned results might
be nothing more than a reflection of the interdependence of
previously used methods. This new technique by Hill, Caudell
and Rosenwald (75) utilizes the velocity amplitude of the
oscillation in conjunction with a more general treatment of
acoustic wave propagation (72) to estimate temperature perturbations, and provides results for the base of the photosphere
which differ significantly from those previously obtained. A
discussion of the relevant features of the treatment of acoustic waves may be found in §5.

One of the paradoxes discussed in §3.2.2 concerns the
values for temperature fluctuation amplitude obtained by
Livingston, Milkey and Slaughter (48) in their 1976 study of
global oscillations. The paradox is generated by a comparison of the results obtained by Livingston, Milkey and Slaughter with those obtained by Hill, Caudell and Rosenwald (71)
based upon an analysis of the observations by Brown, Stebbins
and Hill (42). A possible source of this discrepancy is in the
use by Livingston, Milkey and Slaughter of a theoretical
treatment which considers the solar atmosphere to be a static
system in the evaluation of the relationship of intensity
changes to temperature changes. Hill, Caudell and Rosenwald
(75) have noted that such a treatment may not be applicable
to an oscillating system. Further, the simultaneous observations of fluctuations in continuum brightness and velocity
amplitudes by Tanenbaum et al. (74), coupled with the theoretical work of Hill, Rosenwald and Caudell (72), permit determination of the temperature sensitivity of spectral line
intensity without using line formation theory. Proof of the
validity of a fundamental assumption of this work--the charac-
terization of the five-minute oscillation as a p mode oscilla-
tion--would confirm the usefulness of this method in the more
direct inference of temperature amplitudes. The solution
subsequently obtained is shown in Figure 6.1.

Using the solutions in Figure 6.1 and the observations of
Tanenbaum et al. (74) and Leighton, Noyes and Simon (73), it
was found that the temperature sensitivities of certain spectral lines and of the continuum were actually below those
obtained by other methods (75). In particular it was noted

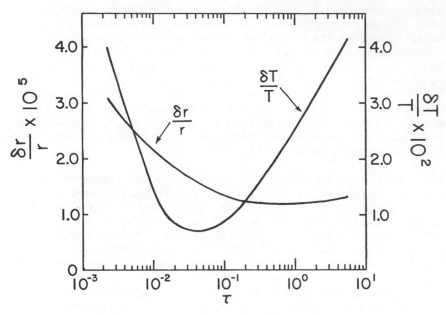

Fig. 6.1. The solutions of δr/r and δT/T for a period of five
minutes as a function of optical depth in the photosphere.   These solutions are linear combinations of
the two appropriate independent solutions to the
wave equation.   The constants of integration were
determined from simultaneous observations of velocity and intensity fluctuations by Tanenbaum et al.
(74).

that the temperature sensitivity of the chromospheric line
Ca II K_{232} was reduced and it was also suggested that it may
be reduced for the photospheric line CI 5380 used by Livingston, Milkey and Slaughter. The latter suggestion was subsequently confirmed by Hill, Livingston and Caudell (76).

The significantly reduced temperature sensitivity reported
in this new work results directly from the strong increase in
$\delta T/T$ at the base of the photosphere (cf., Figure 6.1 near
$\tau = 1$). It has been observed by Hill, Livingston and Caudell
(76) that this sharp increase in $\delta T/T$ is a direct manifestation of the κ-mechanism (5) operating in the hydrogen ionization zone.

Hill, Livingston and Caudell also suggest that the usual
criticisms of such a sharp increase in the value of $\delta T/T$ are
not applicable at the base of the photosphere. One parameter
that is frequently used to characterize such a system is
damping time, i.e., the e-folding time for the decay of a
disturbance. The radiative damping time for the base of the
photosphere most commonly used is that obtained by Spiegel
(87). The magnitude of this time is about 10 sec, a time
quite short when compared to a 5 min period of oscillation.

However, since opacity is very sensitive to temperature
at the base of the photosphere (a fact which was not considered by Spiegel), this fact not only alters, for a p mode
oscillation, the ability of a perturbation to decay away, but
even changes the sign of the process from decay to growth.
In other words, the large increase of opacity with temperature
diminishes the flow of radiation upon compression and enhances
the local rate of temperature increase. Furthermore, the
temperature dependence of the opacity and the temperature
gradient in the photosphere combine to produce only small
changes in radiation intensity. For example, a temperature
increase will bring about an opacity increase which in turn
decreases the depth to which one sees. On the other hand,
the equilibrium temperature varies inversely with elevation
in the atmosphere. Thus, the net observable change in
temperature is the sum of these two effects, where strong
cancellation is possible.

It was concluded by Hill, Livingston and Caudell that
the presumption of small temperature perturbations at the base
of the photosphere for p mode oscillations with 5 min periods
excludes from consideration the current models of the photosphere which place the hydrogen ionization zone near $\tau = 1$
and/or use of the κ-mechanism as a functioning process in

variable stars. This conception clearly does not reflect actual conditions.

The discovery of a large temperature fluctuation associated with solar oscillation at the base of the photosphere is a quite significant result, and represents the second such finding to be obtained through solar seismology. The far-ranging implications of this discovery are explored in the following sections; an initial example of these implications is provided in the role of temperature perturbations in the paradoxes discussed in §3.2.2.

6.3. Evidence for Nonlinear Effects

Analysis of solar p modes has traditionally involved the use of linearized conservation equations. Examples are found in (14), (72) and (83). There are regions in the solar atmosphere where the neglect of second order effects may still permit a good approximation; however, at the base of the photosphere, where $\delta T/T$ becomes large, and in the chromosphere, where the local wave velocity may become large, second order effects can play an important role, considering the large number of excited five-minute modes and the effects of interference among the modes.

A relevant question concerns the existence of observational evidence to support either the exclusive use of linear theory or the inclusion of nonlinear effects. For example, two properties of the five-minute oscillation which suggest that nonlinear effects are present are associated with saturation of the local velocity of the five-minute oscillation and with frequency doubling in the chromosphere. A well-known physical property of nonlinear systems is harmonic generation by a driving signal. In this case, the five-minute oscillation would be considered the driving signal and evidence for the generation of harmonics may exist in the observations shown in Figure 6.2. This figure, taken from the review paper of Noyes (21), shows that significant power appears near a period of 2.5 min in the chromosphere but not in the photosphere. Such harmonic generation would represent a considerable transfer of energy away from the five-minute oscillation. This subsequently requires a reduction in the rate that the velocity increases with height in this oscillation from that predicted by the linear theory. The growth of velocity with height is observed to saturate in the chromosphere, the region where harmonic generation may be present; alternatively, predictions derived from the linear theory indicate an exponential growth rate with a scale height approximately twice that of density.

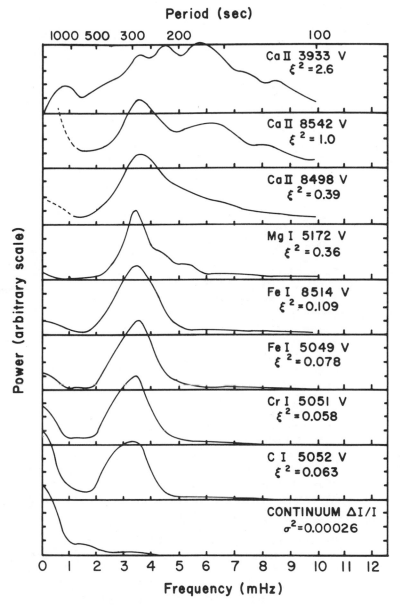

Fig. 6.2. Power spectra of the velocity field for spectral
lines of varying strength and for the intensity
fluctuations in the continuum, as shown by Noyes
(21). The strength of the line increases from bottom
to top of figure as does height of formation in the
atmosphere. All spectra are normalized to maximum
power of unity. The mean square of the velocities
is given by ξ^2 in km^2/sec^2.

Although these two phenomena may be interpreted differ-
ently, the presence of nonlinear processes is ceratinly
possible. The examination of these possibilities is currently
underway and, as pointed out in §7.1.1 and §7.1.2, the study
of the solar atmosphere may require techniques usually
reserved for nonlinear pulsation theory.

6.4. New Constraints on the Convection Zone

One of the topics relevant to a discussion of stellar
pulsation is that of time scales (5). One time scale of interest is that given by the period-mean density relation for the
fundamental radial mode. It is customary to write this relation in the form

$$\Pi(\bar{\rho})^{\frac{1}{2}} = \frac{2}{[\frac{4}{3}\pi G]^{\frac{1}{2}}} \quad , \tag{6.2}$$

where Π is the pulsation period, $\bar{\rho}$ is the mean density of the
star, and G is the constant of gravitation. This period-mean
density relation may be interpreted as illustrating the
insensitivity of the acoustic mode spectrum to details of the
stellar structure. It may also appear that the constraint
on surface composition provided by analysis of the sun's
optical spectrum, coupled with the constraints provided by
solar luminosity, age, and radius, would completely determine
all of the solar model parameters and provide only one accept-
able theoretical acoustical spectrum. These implications may
not be totally justified, however. A more accurate expression
for the period-mean density relation demonstrates that the
right-hand side of equation (6.2) does depend upon the stellar
model (see (5)). In response to the latter assumption, it
was noted by Iben (58) that the uncertainties in surface
opacity and the crudeness of our treatment of convective flow
make both the surface composition and the radius constraints
rather weak for models of the solar interior.

Iben (58) examined the dependence of the theoretical
spectrum of eigenfrequencies on the treatment of convection
near the model surface. The study also noted the higher
sensitivity of the period of the fundamental mode to the convection zone treatment in comparison with higher order modes.
Through such analysis, Iben concluded that important constraints could be placed upon the structure of the convection
zone by studying the acoustic mode spectrum.

However, the initial results from seismic sounding of
the convection zone were due not to the properties of low
order p modes but rather to the five-minute oscillatory mode.
A comparison of the theoretical eigenfrequencies for the

five-minute oscillatory mode (14) and the concentration of power in the k-ω plots by Deubner (34), an example of which is shown in Figure 2.1, exhibited a provocative agreement. It is apparent, however, that despite the general agreement shown, the theoretical eigenfrequencies were all systematically high.

These systematic differences were examined by Gough (88) and he estimated that in order to make the observations and theory agree, the depth of the convection zone must be approximately 50% greater than is usually assumed. These systematic differences were also studied theoretically in some detail by Ulrich and Rhodes (83). Basing their interpretation of k-ω plots upon this analysis, Rhodes, Ulrich and Simon (36) formulated several new constraints which could be placed on the outer envelope of the sun. For example, when including the constraints imposed by the surface abundance of lithium, they concluded that $0.62 < r_e/R_\odot < 0.75$, where r_e is the radius at the base of the convection zone, a conclusion similar to that arived at by Gough (88). Another constraint, involving the metal index Z, invalidates the viability of low Z solar models, which have lower neutrino yields and are thus relevant to the neutrino paradox.

These new results demonstrate the potential applications of this area of study. Further work is necessary, however, before strong conclusions can be drawn. The lack of agreement between the new theoretical results and observations at the shorter wavenumbers indicates the presence of some inadequacy in the theoretical treatment. It is also of interest to consider the mixing length theory of convection used in the analysis by Ulrich and Rhodes (83), who stated without qualification or support that uncertainties in this theory will have no effect on the eigenfrequencies. If important constraints are going to be inferred from this type of analysis and applied to the sun, this point should be further elucidated.

6.5. A Characteristic Horizontal Scale for the Five-Minute Oscillatory Mode

One of the first properties of the solar five-minute oscillation to be recognized was the small scale of the velocity field. As noted in §2.1, the scale is less than 5000 km. The existence of such a scale fits naturally into the p mode interpretation of this oscillatory motion, from the actual scale one can infer the maximum ℓ with which modes are excited. The discovery of a second, larger, horizontal scale for the five-minute mode would probably have equally important

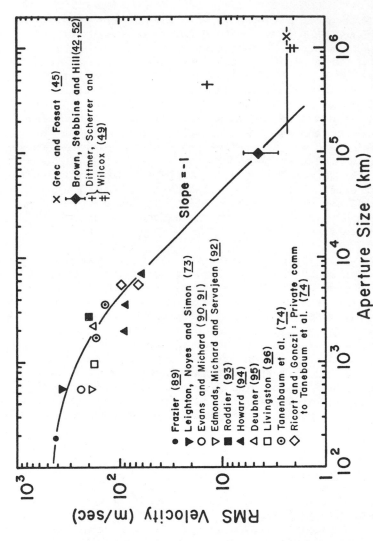

Fig. 6.3. The root mean square amplitude of the velocity observed for the five-minute oscillatory mode as a function of aperture size. Aperture size is the length of a square aperture; circular apertures were reduced to square areas of equivalent area. Consult §6.5.1 for description of treatment of observations by Brown, Stebbins and Hill (42,52) and Dittmer. Scherrer and Wilcox (49).

implications, and observational evidence for such a larger scale does appear to be developing.

6.5.1. <u>Oscillatory Power as a Function of Detector Geometry</u>. Evidence for a larger scale was initially discovered in the course of a study of observed power for the five-minute mode as a function of the detector aperture. In particular, the power spectrum obtained by Brown, Stebbins and Hill (<u>42,52</u>) at a period of five minutes was consistent with velocity observations using small apertures, but not with the power associated with large apertures. In Figure 6.3 the rms velocity obtained by various investigators is plotted as a function of aperture size. Using the procedure of Tanenbaum <u>et al</u>. (<u>74</u>), circular and rectangular apertures were reduced to square apertures of equivalent areas, thereby expressing aperture size in each case as the side length of a square aperture. Figure 6.3 is identical to that of Tanenbaum <u>et al</u>. (<u>74</u>) with the exception of three additional points for large apertures.

At a period of five minutes, the power spectrum of Brown, Stebbins and Hill (<u>42,52</u>) does not exhibit the structure which traditionally characterizes the five-minute oscillation. However, relatively broad peaks do appear in the spectrum at 2.7 mHz, 3.2 mHz, 3.8 mHz, and possibly 4.4 mHz, corrsponding reasonably well to the low wavenumber frequencies at which power is concentrated in the k-ω plot of Deubner (<u>35</u>) (cf. Figure 2.1). Recognizing that the effective aperture of the edge definition technique used by Brown, Stebbins and Hill (<u>42,52</u>) is in fact a spatial filter, the limitations placed upon the permissible values of ℓ are evident. The sensitivity, as shown in §4.2.3, is reduced by $1/\sqrt{2}$ for ℓ approximately 45, corresponding to a wavenumber cutoff at 0.064 (Mm)$^{-1}$. The following analysis is based on the assumption that the power in these peaks is attributable to five-minute modes. This power is estimated to be 370 ± 150 (milli-arcsec)^2; however, in the event that this structure should prove to originate in some other source, 370 ± 150 (milli-arcsec)^2 may be considered to be an estimated upper limit of power.

The conversion of power in diameter measurements to velocity power produces identical results for either coherent or incoherent oscillatory motion of the opposite edges of the sun used in the diameter measurements. This similarity is due to the fact that, for coherent motion, the parity of the spherical harmonic, Y_ℓ^m, permits the detection of only even ℓ modes. The remaining factor necessary to complete the conversion is the determination of the sensitivity of SCLERA type measurements to oscillations. This was accomplished through the use of solutions for $\delta r/r$ and the enhancement factors,

$\Delta R/\delta r$, given by Hill, Rosenwald and Caudell (72), where ΔR is the apparent change in edge location produced by the oscillation. The result, shown as a solid diamond in Figure 6.3, is an rms velocity of 4.4 ± 1.8 m/sec at an optical depth $\tau = 0.02$, having an equivalent aperture of 9.7×10^4 km. The optical depth $\tau = 0.02$ was chosen as typical of the optical depths for which velocity measurements were made.

The whole disk velocity measurements of Grec and Fossat (45) were used to obtain the point indicated as a cross in Figure 6.3 for an aperture of 1.24×10^6 km. Because these observations were quite similar to the other velocity observations listed in Figure 6.3, the determination of this point was relatively simple. In this case no renormalization of the observations was required.

An attempt to generate a third new point was made using the differential velocity measurements of Dittmer, Scherrer and Wilcox (49), whose work is discussed in greater detail in §6.3. However, several difficulties were immediately encountered in association with the determination of equivalent aperture and detector efficiency. Assuming that the oscillations are coherent over the entire area of their detector, a radius of $0.8\ R_\odot$, this question may be resolved through the application of Figure 4.1. As concluded in §4.2.2, the detector efficiency for $\ell \lesssim 12$ is 0.15 while the equivalent aperture for this cutoff in ℓ is 4.4×10^5 km. The rms velocity under this assumption is 14 m/sec and is plotted in Figure 6.3 as a large plus sign.

The observed root mean square velocity in Figure 6.3 is expected to be a monotonically decreasing function of the aperture size; the group of oscillatory modes detected with a larger aperture are a subset of the group of modes detected with a smaller aperture. Obviously the results from the three large apertures plotted in Figure 6.3 do not satisfy this requirement.

There are certainly several places where errors may have entered into the analysis leading to these points. One of the more plausible places where the assumptions may not have been valid concerns the differential measurements of Dittmer, Scherrer and Wilcox (49). It was assumed that the oscillations were coherent over their detector area and the angular dependence of the eigenfunction given by $Y_{\ell m}$. If, on the other hand, the oscillations are incoherent over the detector area or the angular dependence of the eigenfunctions at the surface is not represented with an accuracy of better than 99% by a spherical harmonic, then the Dittmer, Scherrer and Wilcox observations can no longer be properly classified as

differential measurements. In this case, these observations
will probably yield results quite similar to the integrated
signal of a single aperture with radius 0.8 $R_\odot$. Their detector efficiency in this limit would be the same as the other
velocity measurements with the equivalent aperture equal to
0.99 $R_\odot$. This rms velocity is 2.1 m/sec (no renormalization)
and is plotted as a double cross in Figure 6.3.

6.5.2. Anomalous Results for Large Apertures: A Horizontal Scale of Approximately ¼ $R_\odot$. The discussion in §5.6.1
suggests that the observations of Grec and Fossat (45) and
Dittmer, Scherrer and Wilcox (49) may require an interpretation different than that for observations from small equivalent
apertures if one assumes that the p modes are global and that
the angular dependence of the eigenfunctions is described by
the spherical harmonic, Y_ℓ^m, to an accuracy better than 99%.
However, if either or both of these assumptions are relaxed,
these observations are internally self-consistent and are not
inconsistent with other velocity measurements.

The analysis in the preceding section suggests that the
p modes are not global and/or that the proper general solution in θ and ϕ for the oscillation may still be lacking at
the accuracy level beyond 99%. Figure 6.3 indicates, in fact,
that something significant occurs at an equivalent aperture
size of about 1.8×10^5 km or 0.26 $R_\odot$, the intersection of
the horizontal line through the Grec and Fossat observation
and the line passing through the other observations with a
slope of -1. Although its significance has not been clearly
established, it is interesting to note the fact that this
length is typical of depths frequently ascribed to the convection zone.

6.5.3. Interpretation of the Large Horizontal Scale. Both
the global and nonglobal interpretations of the five-minute
mode may accommodate the existence of the larger horizontal
scale discussed previously. For either interpretation, the
discovery of such a horizontal scale would have important
implications.

In the framework of the nonglobal interpretation of the
oscillations, this horizontal scale could be considered as
the characteristic horizontal dimension of the solar surface
involved in a particular p mode oscillation. This would
imply the existence of nonradial boundaries separated by this
dimension, which would effectively decouple the different
regions of the sun.

The horizontal scale for a global oscillation may imply that the spherical harmonic does not offer an adequate description of the eigenfunction at the solar surface for accuracies beyond the 99% level. The scale in this case would reflect the distance for which observational results are adequately described by the spherical harmonic. It is generally assumed that the sun is spherically symmetric when eigenfrequencies, eigenfunctions, etc., are obtained for a solar oscillation. The limitations of this assumption are apparent, however, in that the boundary conditions in the solar envelope are not spatially isotropic, as indicated by the presence of coronal holes (see Hundhausen in this volume). This would change not only the amplitude of the two solutions discussed in §5.2 and §6.1, but also their relative magnitudes. Recalling, as stressed in §5, that these observations relate to the surface region of the sun, the importance of this point is realized. In addition, there may also exist internal nonradial features, which are incapable of decoupling regions of the sun to produce nonglobal oscillations, but which could still effect modification of the eigenfunctions at the 1% level.

6.5.4. Implications for Interpretations of k-ω Plots and the 2 h 40 min Oscillation.

It is interesting to note that this horizontal scale may impose a limiting size for the aperture used in observations to generate k-ω plots (cf. §2.1 and §6.4). At this size (256 arcsec) and larger, the power at the low end of the k scale may be seriously contaminated. Rhodes, Ulrich and Simon (36) utilized a square aperture of 256 arcsec. In view of this, it is suggested that caution be exercised when considering their power spectrum at the lowest values of k.

It is also interesting to ask what the implications of such a horizontal scale would be for the interpretation of the observations on the 2 h 40 min oscillations. Since the work of Brookes, Isaak and van der Raay (46) involves whole disk measurements while the studies of Severny, Kotov, and Tsap (47) and Kotov, Severny and Tsap (61) utilize differential measurements with geometry similar to that of Dittmer, Scherrer and Wilcox (49), it is generally assumed that the observed oscillations are g modes that are low order in ℓ. It is apparent that this conclusion is based upon the assumption, that the oscillations are global and that the angular part of the eigenfunction is given by the spherical harmonic to a very high accuracy.

Should the existence of a large horizontal scale for the five-minute mode signal the breakdown of this latter assumption for this mode, it would be reasonable to expect a breakdown of the assumption for the g modes at a similar level of accuracy. Severny, Kotov and Tsap (47) may well be detecting high order g modes, a supposition quite consistent with the findings of Hill and Caudell (43) that the 68 min and 45 min peaks of Brown, Stebbins and Hill (42,52) are g modes with $\ell \approx 45$.

In this subsection, the systematics of the velocity power in the five-minute mode observed by various groups have been examined; it is concluded that there is evidence for a horizontal scale of a quarter of a solar radius. It is also concluded that such a scale cannot be used to argue either for or against global or nonglobal interpretation of the five-minute mode although should the scale exist, it would be a significant new property in either interpretation. It is further noted that serious implications may exist for the value of the k–ω plots at low k for the five-minute mode, due to possible contamination by modes with larger values of k and for the interpretation of the observations on oscillations with periods near 2 h 40 min.

7. Directions for the Future

The first phase in the development of the new field of study centering on solar oscillations has been largely concerned with the problem of confirming that the observed oscillations are in fact both real and of solar origin. This work has been fraught with difficulties emanating from various sources. One of these sources is the quite subtle manner in which the oscillations themselves modify the observable quantities on the solar surface. Another major problem has stemmed directly from our inadequate understanding of the phenomena of solar and stellar oscillations. Considerable progress has been achieved in both areas as indicated by the review of observations in §2 and the review in §6 of the first findings in solar seismology. An implication of the material reviewed in these two sections is that we have just crossed a threshold and that full utilization of these new techniques may lead to new revelations in solar and stellar physics. Suggested refinements of the observational techniques and concomitant developments in theory as perceived from this vantage point in time are examined in the following subsections.

7.1.   Is Modification of Existing Stellar Envelope Models Necessary?[†]

The sharp rise in the solution for $\delta T/T$ shown in Figure 6.1 was at first perceived to be only an interesting curiosity; however, when considered in the context of the outer envelope, it represents a significant perturbation.

The root-mean-square amplitude of the five-minute oscillation in the photosphere is typically

$$\delta r/r \sim 8 \times 10^{-5} \quad , \tag{7.1}$$

suggesting that this oscillation is only a minor perturbation in the solar envelope. However, the error in this interpretation can be demonstrated by a detailed examination of the opacity in the region where $\tau = 1$. When the results of Figure 6.1 from the velocity observed by Tanenbaum et al. (74) are related to the observed rms velocity of 0.4 km/sec for the five-minute mode (cf. Figure 6.2), we have

$$\langle T/T \rangle_{rms} \sim 0.075 \quad . \tag{7.2}$$

Using the opacity tables of Cox and Tabor (97), one finds that for the King IVa mixture, which is typical of the solar atmosphere, $\kappa_T = 10.5$ at $\tau = 1$, where $\kappa_T = (\partial \ln \kappa)/(\partial \ln T)$. Thus, in linear pulsation theory and neglecting density fluctuations,

$$\left(\frac{\delta \kappa}{\kappa}\right) \simeq \kappa_T \frac{\delta T}{T} \quad , \tag{7.3}$$

yielding

$$\left(\frac{\delta \kappa}{\kappa}\right)_{rms} \simeq 0.78 \quad . \tag{7.4}$$

This does not represent a small perturbation, and, when taken with the large nonlinear dependence of opacity on temperature, it signals a breakdown of linear theory at the base of the photosphere for coherent phenomena. The situation becomes further complicated deeper in the hydrogen ionization zone.

The existence of a region in the solar envelope in which linear theory is inadequate has serious implications that extend beyond the mere complication of pulsation calculations. Importantly, the mean opacity required to generate an equilibrium model of the envelope will not be adequately given by

[†]Contributed by: H.A. Hill, Dept. of Physics, U. of Arizona
                 R.D. Rosenwald, Steward Obs. U. of Arizona
                 T.P. Caudell, Dept. of Physics, U. of Arizona

the standard opacity tables. These results indicate that the presence of the five-minute oscillation must be considered concurrently with both the generation of the opacity tables and the envelope model. Furthermore, an adequate understanding of the hydrogen ionization zone may require the use of nonlinear pulsation theory.

7.2. On Limiting Processes

As noted in the introduction, the apparent lack of a mechanism to effectively limit the amplitude of solar oscillations has been a source of some concern. The reported solar oscillations have very small amplitudes, unlike those observed in the broad class of variable stars (see §3); the limiting mechanisms which are usually considered in stellar pulsation are not operative at such small amplitudes.

The resolution of this difficulty may be found in the degree to which the hydrogen ionization zone can be modified by the five-minute oscillation. It was observed in §7.1 that if the five-minute oscillations are p modes with no local horizontal boundary conditions, then linear pulsation theory is inadequate in the hydrogen ionization zone for treatment of the coherence phenomena associated with the five-minute oscillation. Accordingly, our concept of this zone may require modification. This local nonlinear behavior on the sun is a phenomenon which has been previously observed; for example, it occurs on a global scale for Cepheid variables. However, the overdriving of the hydrogen ionization zone is usually achieved by one or two pulsation modes of large amplitudes in Cepheids, whereas in the sun it is achieved by a very large number ($N \sim 1.6 \times 10^7$) of small amplitude modes. Thus it may be that the limiting amplitude of the five-minute mode would be due not to a highly nonlinear dissipating mechanism but rather to the onset of nonlinear effects in the driving region.

The foregoing estimate of the number of modes is obtained by using the velocity amplitude of an individual mode (36) and the rms velocity for the five-minute mode. This velocity amplitude of each mode at $\tau \sim 0.1$ is approximately 10 cm/sec. Each individual mode has an amplitude much too small to seriously affect the hydrogen ionization zone. However, at any given time and surface location, there would be on the average $N^{\frac{1}{2}}$ states coherent with a lifetime of $\sim 1/\Delta\nu$, where $\Delta\nu$ is the width of the observed power spectrum. This coherence is responsible for the apparent local quality of the five-minute oscillation and for the overdriving of the hydrogen ionization zone. Phenomena analogous to this coherence in the solar envelope are found in other areas of physics.

A coherence phenomenon of this type would assist not only in providing a limiting mechanism for small amplitude solar oscillations but would also serve to counter the arguments that coherence effects are not important in stars. For example, it would be a viable mechanism for the locking together of various modes of oscillation required by Wolff (20) in his recent interpretation of the sunspot cycle.

7.3. Application of Seismic Sounding to the Study of Spectral Lines†

The correlation between the amplitude of the oscillatory motion and the strength of a spectral line (21) was discovered quite early in the work on the five-minute oscillation. A more interesting correlation considered by Noyes (21) was the one existing between the amplitude of oscillatory motion and the heights of formation of absorption lines: the amplitude increased as the height of formation increased (see Figure 6.2). The motivation for examining the latter correlation from that time to the present was the need for a diagnostic tool with which to study the five-minute oscillation.

The development of this diagnostic tool has been difficult at best, primarily because of the complexities associated with the determination of appropriate heights of formation of the absorption lines. However, the observational efforts to obtain power distributions as a function of the horizontal wavenumber, k, and the frequency of oscillation have been quite successful (34,35,36). An example of such a two-dimensional power spectrum in k and ω is shown in Figure 2.1. As noted in §2.1, this type of work strongly supports the interpretation of the five-minute oscillation as a global p mode oscillation because of the good agreement between the predicted and observed structure in these plots.

If the p mode identification is a reasonably accurate interpretation of the five-minute oscillation, it may be productive, as recently noted by Hill, Caudell and Rosenwald (75), to consider revision of the goals in the studies of the five-minute oscillation using the properties of spectral lines. That is, the oscillatory power in terms of velocity and intensity may be used as a diagnostic tool in the study of spectral lines, both as a function of position in the profile of a spectral line and as it varies from spectral line to spectral line. This particular relationship of the theory of line

---

†Contributed by: H.A. Hill, Dept. of Physics, U. of Arizona
R.T. Stebbins, Sac. Peak Obs., New Mexico
R. D. Rosenwald, Steward Obs., U. of Arizona

formation to the theory of solar oscillation in the photosphere would appear to be the better pedagogical one; the
analysis required to determine the properties of the eigenfunctions in the photosphere appears at this time to be much
simpler than that required to predict the manifestation of the
oscillations in the spectral lines.

It was observed by Parnell and Beckers (98) that the
usual arguments concerning the contribution function of spectral lines are not adequate when applied to such phenomena as
the depth variation of solar velocity fields. They pointed
out that the effective depth of formation of absorption lines
depends on the velocity distribution itself. The scope of
this work was extended by Beckers and Milkey (99), who introduced the so-called "Line Intensity Response Function" (henceforth called *response function*), of solar spectral lines to
changes in physical conditions along the line of sight.

The response function originally derived by Beckers and
Milkey can be applied in essentially the same form in the use
of the five-minute oscillation as a probe of spectral lines;
it need only be generalized to include changes in the source
function S and to allow for non-LTE conditions. The general-
ization leads to the following expression for the response
function, $RF(\tau_\lambda)$; at the optical depth τ_λ and wavelength λ:

$$RF(\tau_\lambda) = e^{-\tau} \left\{ \delta S - \frac{\delta \kappa}{\kappa} [I(\tau) - S(\tau)] \right\} \quad , \qquad (7.5)$$

where κ is the opacity at wavelength λ, $\delta\kappa$ and δS are the
Eulerian perturbations in κ and S, respectively, and I is the
outward directed intensity at optical depth τ. The subscript
on τ has been omitted in order to simplify the notation on
the right-hand side of the equation. Note that when studying
observables associated with the five-minute oscillation, the
Eulerian perturbations must be obtained using the appropriate eigenfunctions for the oscillation. As an illustration,
the general solutions in the photosphere are available from
the work of Hill, Rosenwald and Caudell (72).

The two observables associated with the five-minute
oscillation are oscillatory power in velocity and intensity.
The intensity amplitude is easily obtained from equation (7.5)
as

$$\delta I_\lambda = \int_0^\infty RF(\tau_\lambda) \, d\tau_\lambda \ . \qquad (7.6)$$

The velocity amplitude v_λ is given by

$$v_\lambda = \frac{1}{(dI/d\lambda)} \int_0^\infty \omega r \left(\frac{\delta r}{r}\right) e^{-\tau} \frac{\partial \ln \kappa}{\partial \lambda} [I_\lambda(\tau) - S(\tau)] \, d\tau, (7.7)$$

where ω is the angular frequency of the oscillation, r is the solar radius, and $\delta r/r$ is the eigenfunction for the local fractional change in the radius of the sun.

Notice that due to the approximate symmetry of κ about the center of spectral lines, the observable v_λ is primarily sensitive to the $\delta\kappa/\kappa$ term in $RF(\tau_\lambda)$ while δI_λ is sensitive to both the δS and $\delta\kappa/\kappa$ terms. Thus, simultaneous observation of both quantities should permit the separation of their contributions to the response function.

Another quantity which can be extracted directly from the velocity and intensity observations is the effective optical depth, τ_λ^{eff}. Parnell and Beckers (98) quite appropriately defined effective optical depth as that optical depth at which the true velocity $v(\tau_\lambda^{eff})$ corresponds to the observed velocity v, i.e.,

$$v(\tau_\lambda^{eff}) \equiv v_\lambda , \qquad (7.8)$$

where v_λ is given by equation (7.7).

An observational program is currently in progress to study the observables δI_λ and v_λ and the inferred effective optical depth, τ_λ^{eff}, as functions of λ over certain spectral lines. The lines chosen are considered to be well understood and are observationally "clean". The initial emphasis of the work is the evaluation of prospects for the use of five-minute oscillations as a probe of spectral line formation.

7.4. Test of Nonlinear Theory in the Low Photosphere and the Chromosphere

The manifestation of nonlinear effects in the oscillatory motions of the chromosphere is important with regard to heating of the corona and chromosphere. A detailed understanding of the heating mechanism has proven elusive, in part because of the complexity of the nonlinear effects involved and the lack of a quantitative method with which to study such phenomena.

Should the evidence for nonlinear effects as discussed in §6.3 in fact be valid, the value of the five-minute oscillation as a probe in another area is suggested: it would serve as a tool to quantitatively test various theories dealing with nonlinear processes that may be operative.

The discussion of §7.3 did not distinguish between a velocity field of small amplitude and one of large amplitude. It was tacitly assumed that linear theory was adequate and accordingly the equations given were obtained for such an approximation. However, with the five-minute oscillation there is another variable at our disposal. This variable is the number of modes that are coherent at a particular time and surface location, i.e., the magnitude of the velocity amplitude at a certain optical depth. The use of this variable as a parameter in observations of the type outlined in §7.3 would not only be pertinent to the quantitative study of nonlinear effects, but could contribute to a better definition of the boundary conditions that are relevant to the five-minute oscillation.

The acquisition of this kind of information is an additional aim of the observational program discussed in §7.3. As an adjunct to the implementation of this program, the development of the analytical tools necessary for interpreting the observations is underway along lines similar to those required in calculations dealing with nonlinear stellar pulsation.

7.5. Probing the Boundary Conditions

The boundary conditions in stellar pulsation theory have traditionally been set and predictions made with regard to such quantities as observable surface velocities and brightness fluctuations. For solar oscillations, it has been demonstrated that the roles of theory and observation can be fruitfully reversed to give detailed information about boundary conditions not previously available. Had there been no surprises in the first results of work where the roles were reversed (see §6.1), future work on boundary conditions would be considered a secondary program. However, the first results are quite inconsistent with those anticipated, making future work in this area fundamental; these unforeseen results indicate the incompleteness of our understanding of oscillations in the chromosphere and corona and consequently, the possible inadequacy of our current treatment of heating in these two regions by wave phenomena.

The intercomparisons discussed in §6.1 have, at this point in time, placed constraints on the relative amplitudes of the two solutions of the wave equation (see §5). It may be possible by further refinement of the observational techniques to carry these intercomparisons to the level where more quantitative statements can be made about the amplitudes, particularly their dependence on the period of oscillation.

7.6. Probing the Solar Interior

The possibility of observational detection of global solar oscillations was announced (3) not long after the discovery of a lower than expected solar neutrino flux which indicated a major problem existed in our understanding of the solar interior (see the discussion by Davis in this volume). It was recognized at the outset of work on solar oscillations that if sufficient detailed information could be obtained about a large number of oscillatory modes, we would then have an independent probe of the solar interior similar to that used by seismologists for study of the earth's interior.

Work on solar oscillations in the immediate future will primarily be concerned with the refinement of observational techniques in order to permit the resolution of individual modes of oscillation, both with respect to frequency and spatial properties. Such resolution will allow the modes of the sun to be classified in much the same way that excited states of the atom and atomic nuclei are presently classified. The analysis by Hill and Caudell (43) suggests that several modes have already been resolved in frequency. The results of work on spatial filter functions such as that discussed in Appendix A are encouraging since they suggest that the spatial resolution of the modes may also be anticipated in the near future. Spectroscopic analysis of the sun's oscillatory modes remains one of the primary and most challenging goals in the future work.

7.7. Prospects for the Observational Study of the Internal Solar Rotation[†]

The study of solar rotation is the best established of those disciplines dealing with large scale motion. However, as has been noted by Beckers and Canfield (22), one striking aspect of this area of study is the large variation in the values reported for solar rotational velocity by different investigators - variations which are much larger than probable errors. An indication of the magnitude of such variation may be seen by comparing the higher rotational velocity of 14.6°/day observed for large, long-lived magnetic features such as sunspots and coronal structures with the lower rate of 13.5°/day associated with the quiet, undisturbed solar atmosphere. In addition, the apparent solar rotation has been observed to fluctuate by as much as a few percent within a period of a few days time (100); (see (101) for further discussion of solar rotation and its implication for the overall structure of the sun).

[†]Contributed by: H.A. Hill, Dept. of Physics, U. of Arizona
                   R.T. Stebbins, Sac. Peak Obs. Sunspot, N.M.
                   T.P. Caudell, Dept. of Physics, U. of Arizona

Such examples of the variation in rotational velocity demonstrate the importance of rotation studies. They are also suggestive of a need for a means by which the internal rotation of the sun might be studied more directly. It has been recognized for some time (cf. (102), §82) that the internal rotation of a star does make small perturbations in the eigenfrequencies of oscillation. It now appears that these very small perturbations may be a means by which a more direct study of internal rotation will be possible.

If the sun were not rotating there would be a $(2\ell+1)$-fold degeneracy in the eigenfrequency, where ℓ is the principle order number of the spherical harmonic. Rotation removes this degeneracy and for the case of uniform rotation, the eigenfrequencies $\omega_{n\ell m}$ take the form (103)

$$\omega_{n\ell m} = \omega_{n\ell} + mc_{n\ell}\Omega , \quad -\ell \leq m \leq \ell, \qquad (7.9)$$

as seen in a frame rotating with the angular velocity Ω of the sun; n represents the radial order of the eigenfunction, m the azimuthal order number of the spherical harmonic, and $c_{n\ell}$ a factor which depends on the eigenfunction. A formula of the same form holds for a non-uniformly rotating star (67), where Ω becomes the appropriately averaged angular velocity. Viewed from an intertial frame, these oscillations have frequencies

$$\omega'_{n\ell m} = \omega_{n\ell} + m\beta_{n\ell}\overline{\Omega} , \qquad (7.10)$$

where $\beta_{n\ell} = 1 - c_{n\ell}$.

The magnitude of the m coefficient in equation (7.10), $\beta_{n\ell}\overline{\Omega}/2\pi$, is probably ≈ 0.4 µHz if the internal rotation is similar to the surface rotation, an assumption that is supported by observations on solar oblateness (54). This frequency splitting by the solar rotation, i.e. the m dependent term of equation (7.10), is obviously quite small and has for the most part been considered unimportant.

However, several interesting phenomena have recently been noted which might result from such a small splitting. For instance, the size and systematics anticipated for this splitting have led Wolff (20) to suggest that it may be possible to make different oscillatory modes oscillate at identical frequencies with quite profound consequences. In another instance, the fluctuations in the Princeton solar oblateness data have been interpreted as resulting from this rotational splitting (66,67) Should this interpretation be correct, a new measure of solar rotation will have been found. It has also been recognized recently that this rotational splitting

(equation 7.10) should be directly observable in the study of
solar oscillations (103,104). Because the latter two endeavors
relate directly to a possible measure of internal rotation,
they are discussed in greater detail in the following subsections.

7.7.1. Beating Between Oscillatory Modes: Long Period Fluctuations in the Princeton Solar Oblateness Observations.

The Princeton solar oblateness data (63) have recently been
reanalyzed in an effort to understand the sizable day-to-day
variation in those data (64,65). Much of the variation was
found to be consistent with periodic changes in the apparent
oblateness of the solar image, having an amplitude of about
2×10^{-5} and a synodic period of 12.64 ± 0.12d. Dicke (64)
concluded that the results obtained indicated a distortion of
the solar surface which rotated rigidly with a period of
12.64d. He did, however, note that one alternative hypothesis
which could not be excluded statistically was that the signal
resulted from an inhomogeneous photospheric brightness, i.e.
a non-uniform surface brightness, rotating rigidly with a
period of 12.64d. In addition to pointing out the severity
of the energy and momentum balance conditions, Dicke (64)
observed that, "Also a rapidly rotating brightness distribution is no less heretical than a rapidly rotating shape irregu-
larity."

Gough (66) noted that a rapidly rotating brightness
distribution may not be heretical at all, if it represents
the direct manifestiation of mode splitting discussed in §7.7.
The observed long period rotating signal would then be a manifestation of beats between modes of the same n and ℓ but
different m. Using this interpretation of the fluctuating
component of the Princeton oblateness signal, Gough (67) concluded that the average angular velocity in the solar interior
must be approximately two to four times the equatorial angular
velocity at the solar surface.

The analysis by Gough (67) was limited in its conclusion
primarily due to a lack of spatial information on the relevant
oscillations. However, should his interpretation prove to be
correct, exciting prospects are seen for internal rotation
studies of this nature.

7.7.2. Broadening of the Power Spectrum for the Five-Minute Mode by Rotational Splitting.

In addition to the
generation of beats as discussed in §7.7.1, observation of
rotational splitting of the eigenfrequencies might be accomplished in several additional ways. However, it must be
realized that the success of a particular observational program will depend in large measure upon the stability of a

particular mode with time. This stability will be determined by, among other things, the primary excitation or driving mechanism, the primary dissipation mechanism, and whether or not the oscillations of interest represent oscillations of the whole sun. The latter criterion relates to the degree that oscillations in one region of the sun are the same as those oscillations found in adjacent regions, a phenomenon that may be addressed by the horizontal scale discussed in §6.5. However, a negative result with regard to detecting rotational splitting may in itself prove to be quite interesting in the ensuing constraints on these three points.

The three groups of solar oscillations discussed in §§2.1, 2.2, and 2.3 (the oscillations with periods of five minutes, between five minutes and one hour, and near $2^h 40^m$) are all candidates for use in the study of rotational splitting, assuming of course that the more controversial issues concerning these observations are resolved in a positive sense. Work is now in progress at SCLERA to detect rotational splitting in the longer period modes and at the Sacramento Peak Observatory to detect the effects of rotation on the five-minute mode.

Two different approaches to the detection of the effects of rotation should be discussed. The first, which is appropriate for the five-minute mode, essentially identifies the effects of the rotation on the power spectrum of the oscillation. The second approach, discussed in §7.7.3, identifies the individual modes through the appropriate choice of spatial and temporal filtering.

The study of the effects of rotational splitting on the power spectrum of the five-minute mode is attractive in that it does not require any particular stability of eigenmodes. This is an important point since a particular five-minute mode may not exhibit oscillatory motion which is coherent over a very long period of time. Supplementally, the discussion of this technique will show that the work on the k-ω plots (cf. §2.1) may in some instances be compromised *a priori* by rotational effects introduced by insufficient tracking accuracy of the observing instrument. In any case, further refinement of this type of work will certainly be hindered in some cases by the limitations inherent in existing telescope tracking mechanisms.

Using equation (7.10), the velocity amplitude may be written as

$$A(\theta,\phi,t) = \sum_{n,\ell,m} A_{n\ell m} P_\ell^m(\theta,\phi) \exp\left[i(\omega_{n\ell}+m\beta_{n\ell}\overline{\Omega})t + im\phi\right], \quad (7.11)$$

where $A_{n\ell m}$ is the amplitude of the individual eigenmode, P_ℓ^m

is the associated Legendre polynomial of the first kind, and θ and ϕ are the usual spherical coordinates. For the five-minute mode, $(1-\beta_{n\ell}) \ll 1$ since $(1-\beta_{n\ell}) \propto 1/\ell(\ell+1)$ (cf. <u>67</u>,<u>102</u>) and a typical ℓ is of the order of 400. For m = 400 and assuming that $\overline{\Omega}/2\pi = 0.4$ µHz for the five-minute oscillation, we have

$$\frac{m\overline{\Omega}}{2\pi} = 0.16 \text{ mHz} \quad . \qquad (7.12)$$

This rotation term is 5% of the unperturbed frequency $\omega_{n\ell}$.

This analysis shows that the rotational terms can make a significant contribution to $\omega_{n\ell}$ and immediately suggests the type of observation for the measurement of $\overline{\Omega}$. A set of power spectra may be obtained in which, for each individual spectrum, the telescope has been programmed to introduce a predetermined tracking error that would make the sun appear to rotate at a different rate as seen by the detector. Let the tracking error, $(d\phi/dt)$, which yields the power spectrum having the minimum width be denoted by $(d\phi/dt)_{min}$. It becomes apparent in this instance that

$$\overline{\Omega} = -(d\phi/dt)_{min} \quad . \qquad (7.13)$$

At

$$\overline{\Omega}/2\pi = 0.4 \text{ µHz} \quad ,$$

$$(d\phi/dt)_{min} = -9 \text{ arcsec/hr} \quad . \qquad (7.14)$$

The random tracking errors in the telescope obviously enter in the same way as those due to solar rotation, the effects of errors in θ being similar to those in ϕ. The power spectra can thus be significantly altered by these errors. Equation (7.14) indicates a scale by which the magnitude of such errors can be judged. A random fluctuating error with a rms velocity of 9 arcsec/hr would be expected to increase the width of the power spectrum by about 5%. Not only does this give a criterion by which to judge tracking errors, but it also suggests that some of the differences between reported power spectra may be due to such an effect. Of a more serious consequence are the significant systematic shifts which may result in the observed structure of the k-ω plots for large ℓ. These shifts would be produced by changes in the observed frequency of oscillation. The error introduced by the shifts depends upon the manner in which the k-ω plots are obtained and the degree of isotropy of the oscillatory power in m.

7.7.3. <u>Detection of Rotational Splitting by Spatial and Temporal Filtering</u>. Spatial and temporal filtering, sometimes considered the most direct approach to the study of rotational splitting, may in fact prove to be the most difficult, as noted in §7.7.2. However, a successful execution of this type of observation for the five-minute mode and the longer period modes would permit the study of solar internal rotation as a function of radius. The possibility that this represents a diagnostic probe for use in the study of differential internal rotation has led to the initiation of such an observational program at SCLERA.

The sensitivity of rotational splitting to internal rotation, and hence the radial differential rotation, is most easily obtained by examination of the $m\beta_{n\ell}\overline{\Omega}$ term of equation (7.10). For the case where Ω is a function of only the radial coordinate, Gough (67) has shown that

$$\beta_{n\ell}\overline{\Omega} = \frac{\displaystyle\int_0^{R_{\odot}} \rho\Omega \left\{ (\xi_{n\ell}-\eta_{n\ell})^2 + [\ell(\ell+1) - 2]\eta_{n\ell}^2 \right\} r^2 \, dr}{\displaystyle\int_0^{R_{\odot}} \rho \left\{ \xi_{n\ell}^2 + \ell(\ell+1)\eta_{n\ell}^2 \right\} r^2 \, dr}, \quad (7.15)$$

where ρ is the density, $R_{\odot}$ is the radius of the sun, and the remaining quantities are defined by the components of the displacement amplitude for a normal mode of a nonrotating star,

$$\left\{ \xi_{n\ell}(r) \, P_{\ell}^m (\cos\theta) \, e^{im\phi}, \; \eta_{n\ell}(r) \, \frac{d}{d\theta} P_{\ell}^m (\cos\theta) \, e^{im\phi}, \right.$$
$$\left. im \, \eta_{n\ell}(r) \, \frac{1}{\sin\theta} P_{\ell}^m (\cos\theta) \, e^{im\phi} \right\} \quad . \qquad (7.16)$$

Thus $\beta_{n\ell}\overline{\Omega}$ is a weighted average of the internal rotation with weighting factors being due to the eigenfunctions. Since the internal region where the $\xi_{n\ell}$ and $\eta_{n\ell}$ remain relatively large depends strongly on the order and class of the modes, it is apparent that the study of rotational splitting as a function of mode order and class can yield rather detailed information about the internal rotation.

Ando and Osaki (14) have estimated the effective depth by plotting $\rho\xi_{n\ell}^2$ for the p_1 mode with $\ell = 200$ as a function of radius. This term makes a significant contribution only for the outer 10^4 km of the sun. Assuming that the contribution of $\eta_{n\ell}$ in this case is negligible, since the frequency greatly exceeds the local critical acoustic frequency (cf. Appendix B of (72)), $\overline{\Omega}$ will in this case reflect the mean rotation rate in the outer 1.4% of the sun.

In contrast to this, low order g modes are effective throughout the sun. Examples of $\xi_{n\ell}$ for low n and ℓ may be found in (6),(8),(10),(106). These works clearly show that $\xi_{n\ell}$ remains large throughout the entire sun. Thus $\overline{\Omega}$ in this case should reflect a mean rotation characteristic of the whole sun. Examples intermediate to these two extremes should be offered by low order p modes.

The design of the appropriate spatial and temporal masks requires a rather careful analysis for the unambiguous interpretation of the observations in terms of rotational splitting. The primary source of complexity in this problem is the necessity of using the spherical harmonic Y_ℓ^m to describe the non-radial character of the eigenfunctions. While the Y_ℓ^m form an orthogonal set over the spherical surface, they are not orthogonal over the observed portion of the surface.

The development of a mathematical formalism that can be conveniently used for such analysis is too lengthy to be presented here and will be given elsewhere. However, the example outlined in §3 for the analysis of the detector sensitivity of Dittmer, Scherrer and Wilcox (49) has proven quite powerful in these problems. The prominent feature of this technique is the incorporation of the finite Fourier transform of the spherical harmonic.

The accuracy to be expected for the rotation rate $\overline{\Omega}$, if rotational splitting is observed, will probably be limited by the accuracy to which the solar image can be tracked. Tracking errors will have much the same effect in these considerations as in power spectrum broadening (see §7.7.2). The edge definition techniques used at SCLERA (70) to locate a point on the solar limb can be utilized fully in this case. The use of this edge definition should not pose limits above a few parts in 10^4 in the accuracy of the measurements.

Preliminary analysis indicates that spatial and temporal filtering can be designed to make possible unambiguous interpretation of rotational effects. If the rotational splitting can be identified, the SCLERA facility should permit observations of the internal rotation with an accuracy of parts in 10^4. The ultimate success of such a program, however, will strongly depend upon the existence of global oscillations and the stability of the modes, both of which remain to be firmly established.

7.8 Extension of Observational Techniques

The single most important advance in observational technique relevant to the study of solar oscillations has been the

introduction of the FFTD described in §3.1.3. Several of the
results that have been obtained using this new technique of
edge definition clearly map out a possible direction for
future developmental work and the refinement of that observational technique. Demonstration of the phase coherency of
longer period oscillations over an extended period of time and
the detection of these oscillations through intensity changes
in the limb darkening functions are two of these results. The
manner in which these discoveries can be utilized are discussed
below.

7.8.1. Phase Coherency. The implications of the phase
coherency found by Brown, Stebbins and Hill (52) and Hill and
Caudell (43) (see §2.2) are that both temporal and spatial
properties are sufficiently well defined so that the individual oscillatory modes can be isolated and classified. Enhanced
frequency resolution can be obtained through Fourier analysis
of several contiguous days of observations instead of isolated
single days. The spatial filtering operator, although simple
in conception, is quite complex in practice. Continuation and
refinement of the work discussed in Appendix A will make the
design of an effective observation program much easier.

7.8.2. Brightness Changes in the Limb Darkening
Function. The demonstration by Hill and Caudell (43) that
longer period oscillations may be detected through brightness
changes in the limb darkening function (see §2.2 and §6.1) can
provide a significant tool for future observational programs;
they should no longer be limited to the measurement of the
solar diameter. By observing other Fourier coefficients of
the limb darkening function in addition to the one used in
the FFTD, i.e. use of a weighting function different from that
given by equation (3.2), it should be possible to observe the
oscillations in the location of the apparent limb at any
particular position angle on the solar disk.

For this new mode of observation to be effective, combinations of the various Fourier coefficients must be available
which are sensitive to oscillations while maintaining a
reduced atmospheric seeing sensitivity comparable to that
found in the FFTD. A mathematical formalism has been developed by Hill et al. (107) which should facilitate the
implementation of such a program. The observational demonstration of the feasibility of this extension of the FFTD is
currently being attempted.

8. Summary

The observational evidence which suggests the existence
of global solar oscillations has been reviewed with the

conclusion that a fairly strong case can be made for the following hypothesis: global oscillations are present in the sun with amplitudes sufficiently large so as to be observable. The most impressive findings which support this hypothesis are: (1) a detailed structure exists in the k-ω diagrams for the five-minute oscillations which agrees with previous model calculations and, (2) the longer period oscillations exhibit moderate coherency over a long period of time. An argument for the existence of global oscillations based upon the observed coherency is a compelling one because of the difficulty in finding a source of oscillations other than global oscillations of the sun that produce periods which are not harmonics of the earth's rotation frequency and whose frequencies are stable for long periods of time.

The existence of oscillatory coherency also bodes well for the future of seismic sounding of the sun's internal structure. In order for seismic sounding to have any significant impact upon the development of solar models, the eigenfrequencies probably need to be observationally known to the 0.01% level. The observed coherency indicates that this should not be difficult to obtain in future studies.

The search for solar oscillations has utilized several observational techniques with a variety of observational parameters. These methods have been reviewed in the context of the differing results that have been reported. In this review, some emphasis was given to the new technique developed at SCLERA for the measurement of solar diameters, since this technique appears to be effective in the detection of the longer period solar oscillations.

A review of the literature reveals a number of discrepancies which arise when intercomparisons are made between measurements of different solar phenomena, each of which should be capable of revealing the presence of global solar oscillations. Possibly, many of these discrepancies originate in the complexities of the intercomparisons themselves. Three of the major problems encountered have been classified and analyzed; these concern the geometry of the detector (where the effects are given by a spatial filter function), the influence of the solar atmosphere upon the detection of solar oscillations, and the changes in spectral lines produced by oscillations.

The majority of the first findings from seismic sounding of the sun have been derived from careful studies of these problems; a major outcome of this work has been the realization that, although seismic sounding promises to be of value to studies of the solar interior, it need not be restricted to that application. Seismic sounding has been found to be a sensitive probe of the outer boundary conditions used in stellar pulsation theory and of the temperature perturbations in the solar envelope.

Many important areas of future study have been suggested by recent work, the more obvious of which have been discussed in some detail. The most important prospects concern the continued study of the boundary condition problem, the development of seismic sounding as a new probe for the study of spectral line formation, the study of non-linear effects produced by oscillations in the outer layers of the sun and their relation to stellar atmospheres, and the study of the internal structure and rotation of the sun. The pursuit of these lines of research promises to be both challenging and rewarding.

Acknowledgements

The author wishes to acknowledge the contributions made in the preparation of this paper by the following individuals: O. R. White, in addition to serving as an editorial referee, suggested the emphasis of a number of important ideas that had received only minor attention in the original draft; R. T. Stebbins provided many valuable suggestions for the organization of the paper; and G. D. Harwood and A. K. Whitehead provided technical editing and writing assistance.

Appendix A: On the Calculation
of Spatial Filter Functions[†]

Knowledge of the properties of the spatial filter function for a particular observation on solar oscillations is a fundamental necessity before any semblance of order can be brought to studies of solar oscillations (see §6.1). This is a consequence of the complexity of the oscillatory modes with regard to their spatial properties and the large variety of detector geometries that can be used. In this appendix, the subject material is introduced by a brief review of the previous treatments of the topic and is followed by a discussion of the spatial properties of the modes. A formalism is then outlined which utilizes a particular spatial representation of the oscillations that generally simplifies the analysis of the filter function.

Examples of the spatial filter functions for some of the simpler cases are available in the literature, i.e. for ℓ, m $\leqslant$ 2 (45,78). Another treatment of this problem may be

---

[†]Contributed by: H.A. Hill, Dept. of Physics, U. of Arizona
R.D. Rosenwald, Steward Obs., U. of Arizona

found in the generation of the three dimensional k-ω plots
for the five-minute oscillation discussed in §2.1. This generation requires observational information in two non-radial
dimensions. There have been two different procedures used by
which these two dimensions are collapsed into a single one for
the k scale of the k-ω plots. Deubner (34) used a long,
narrow detector array to obtain detailed information of the
velocity field in one non-radial dimension and an integral
equation by Uberoi (108) for calculation of the spatial density
in k and ω. Rhodes, Ulrich and Simon (36) obtained detailed
information about the velocity field in one non-radial dimension and integrated over a sizable distance in the second non-
radial dimension. In this case they argued that the relation
between k and ℓ was determined to a high accuracy by the integration in the second dimension.

Both of these treatments have assumed that the solution
to the wave equation at the surface of the sun can be written
as

$$e^{i(k_x x + k_y y)},$$

where x and y are two orthogonal coordinates on the surface
and k_x and k_y are, respectively, the projections of the non-
radial wave numbers onto these coordinates. The derivation
of the integral transform used by Deubner (34) begins with
this assumption in equation (1) of Uberoi (108) Rhodes, Ulrich
and Simon (36) also made the same assumption. But in this
case they argue that the integration in one dimension con-
strains $k_y \ll k_x$, and therefore $k = k_x$, where it has been
assumed that the integration is over the y coordinate. However, the plane wave representation is not an eigenfunction
of the non-radial wave equation and, in fact, it represents
the superposition of many eigenfunctions. Fortunately, it
appears that no significant errors have been introduced by
this simplified treatment to date, but errors may well be
introduced at the next level of refinements in these types of
observations.

The natural coordinates for description of the oscillations are the spherical coordinates whose axis defined by
θ = 0 coincides with that of rotation of the sun. In this
coordinate system, the wave equation is separable, with the
non-radial properties given by the spherical harmonics, Y_ℓ^m.
This spatial representation of the oscillations is used in the
following discussion.

The two observables that have been used in the study of
solar oscillations are changes in radiation intensity δI and
velocity. The observed values of these quantities are given

by the averages

$$\langle \delta I \rangle = \frac{1}{\iint \overline{ds} \cdot \hat{x}} \iint \delta I \; \overline{ds} \cdot \hat{x} \tag{A1}$$

and

$$\langle v \rangle = \frac{1}{\iint \overline{ds} \cdot \hat{x}} \iint \overline{v} \cdot \hat{x} \; \overline{ds} \cdot \hat{x} \;, \tag{A2}$$

where $\hat{x}$ is a unit vector in the direction of the observer, $\overline{ds}$ is the infinitesimal surface area normal to the solar surface, and the integration is over an area defined by the detector geometry. If limb darkening effects are neglected, these two integrals for p modes reduce to

$$\langle \delta I_{\ell m} \rangle = \frac{A^I_{\ell m}}{\iint \overline{ds} \cdot \hat{x}} \iint Y^m_\ell (\theta,\phi) \; \sin^2\theta \; \cos \phi \; d\theta \; d\phi \tag{A3}$$

and

$$\langle v_{\ell m} \rangle = \frac{A^v_{\ell m}}{\iint \overline{ds} \cdot \hat{x}} \iint Y^m_\ell (\theta,\phi) \; \sin^3\theta \; \cos^2\phi \; d\theta \; d\phi \;, \tag{A4}$$

where $A^I_{\ell m}$ and $A^v_{\ell m}$ are the respective amplitudes, i.e.,

$$\delta I_{\ell m} = A^I_{\ell m} \; Y^m_\ell \tag{A5}$$

and

$$\overline{v}_{\ell m} = A^v_{\ell m} \; Y^m_\ell \; \hat{r} \;. \tag{A6}$$

For g modes, equations (A4) and (A6) need to be augmented by the appropriate non-radial components.

Inspection of the integrands in equations (A3) and (A4) shows that the integral that must be evaluated is of the form

$$\iint Y^m_\ell \; \sin^{2+\varepsilon}\theta \; \cos^{1+\varepsilon}\phi \; d\theta \; d\phi \;, \tag{A7}$$

with $\varepsilon = 0$ and 1 for the intensity and velocity measurements respectively.

The neglect of the limb darkening effects introduces only minor errors. However, these can also be taken into account fairly well by the appropriate choice of ε. For this reason the general treatment that follows will not assume any particular value for ε.

The particular spherical coordinate system has been chosen so as to simplify the calculation of the eigenfunctions. However, using this coordinate system leads in general to a

double integration in equation (A7) which is coupled in the
limits of integration. For the whole disk, differential, and
the SCLERA type of measurements, this complication can be
removed to a high degree of accuracy by making the appropriate
rotation of the coordinate system with the rotation operator.
For the spherical harmonic we obtain (cf. (77), §4.1)

$$Y_\ell^m(\theta',\phi') = \sum_{m'} \mathcal{D}_{m'm}^\ell(\alpha,\beta,\gamma)\; Y_\ell^{m'}(\theta,\phi) \quad , \tag{A8}$$

where the primed and unprimed variables are the old and new
coordinate systems respectively, the $\mathcal{D}_{m'm}^\ell(\alpha,\beta,\gamma)$ are the matrix
elements of the finite rotation operator, and α, β and γ are
the Euler angles of the rotation.

The rotation that will decouple the integration for the
whole disk measurements of Grec and Fossat (45) and for the
differential measurements of Dittmer, Scherrer and Wilcox (49)
places the new axis, defined by $\theta = 0$, towards the observer,
i.e. $\alpha = \gamma = 0$ and $\beta = \pi/2$. The integration in ϕ is now
between 0 and 2π, which leads to non-zero terms only for
$m' = 0$. This leads to a considerable simplification in the
matrix elements, with

$$\mathcal{D}_{0m}^\ell\left(0,\frac{\pi}{2},0\right) = \left[\frac{4\pi}{2\ell+1}\right]^{\frac{1}{2}} Y_\ell^m\left(\frac{\pi}{2},0\right) \quad . \tag{A9}$$

There are two different types of rotations which will simplify
the integrations for the SCLERA type of measurements; the
choice will depend on the details of the particular measurements. In this case the matrix elements reduce to a combination of gamma functions and Jacobi polynomials.

The remaining integral in θ involves the associated
Legendre polynomial P_ℓ^m and the window function W for the
detector geometry. A quite powerful way to treat such a problem is to represent the window function as a Fourier series,
subsequently changing the θ integration in equation (A7) to
the Fourier analysis of the associated Legendre polynomial.
The details of this Fourier analysis are too lengthy for this
review and will be given elsewhere. However, the resulting
Fourier coefficients are as follows: for $\ell+m$ and q even,

$$\int_0^\pi P_\ell^m(\cos\theta)\,\sin^\eta\theta\,\cos q\theta\,d\theta = \frac{(-1)^{m+q/2}}{2^{2m+\eta}}\pi\;\frac{\Gamma(m+\eta+1)}{\Gamma(m+1)}\;\frac{\Gamma(\ell+m+1)}{\Gamma(\ell-m+1)}$$

$$\cdot\frac{1}{\Gamma[(m+\eta+q+2)/2]}\;\frac{1}{\Gamma[(m+\eta-q+2)/2]}$$

$$\cdot{}_4F_3\left[\frac{\ell}{2}+\frac{m}{2}+\frac{1}{2},\; -\frac{\ell}{2}+\frac{m}{2},\; \frac{m}{2}+\frac{\eta}{2}+\frac{1}{2},\; \frac{m}{2}+\frac{\eta}{2}+1;\right.$$

$$\left.\frac{m}{2}+\frac{\eta}{2}+\frac{q}{2}+1,\; \frac{m}{2}+\frac{\eta}{2}-\frac{q}{2}+1,\; m+1;\; 1\right] \quad , \tag{A10}$$

where $_4F_3$ is the generalized hypergeometric function (cf. <u>109</u>, §9.14). The sine series terms are identically zero. For odd $\ell+m$ and even q,

$$\int_0^\pi P_\ell^m (\cos\theta) \sin^\eta \theta \sin q\theta \, d\theta = \frac{(-1)^{m+(q/2)+1}}{2^{2m+\eta+1}} \pi \frac{\Gamma(m+\eta+1)}{\Gamma(m+1)}$$

$$\cdot \frac{\Gamma(\ell+m+1)}{\Gamma(\ell-m+1)} \frac{q}{\Gamma(m/2+\eta/2+q/2+3/2)} \frac{1}{\Gamma(m/2+\eta/2-q/2+3/2)}$$

$$\cdot \, _4F_3 \left[\frac{\ell}{2} + \frac{m}{2} + 1, -\frac{\ell}{2} + \frac{m}{2} + \frac{1}{2}, \frac{m}{2} + \frac{\eta}{2} + \frac{1}{2}; \frac{m}{2} + \frac{\eta}{2} + 1; \right.$$

$$\left. \frac{m}{2} + \frac{\eta}{2} + \frac{q}{2} + \frac{3}{2}, \frac{m}{2} + \frac{\eta}{2} - \frac{q}{2} + \frac{3}{2}, m+1; 1 \right] , \tag{A11}$$

with the cosine series identically equal to zero.

These results have been used to examine the spatial filter functions for the whole disk velocity and intensity measurements and for the differential velocity measurements. These results are given and discussed in §4.2.1 and §4.2.2.

The completion of the work on the spatial filter function for the FFTD requires that the solutions for δI found by Hill, Rosenwald and Caudell (<u>72</u>) be multiplied by $e^{im\phi}$, followed by an application of the FFTD algorithm to determine apparent shifts in the location of the limb. The requisite FFTD calculations were performed on the 25 min period solution for δI and the results are given in §4.2.3.

References and Notes

1. Hill, H. A., and Stebbins, R. T. 1974, The Seventh International Conference on General Relativity and Gravitation, Tel-Aviv University, June 23-28.

2. Richard, J.-P. 1975, <u>General Relativity and Gravitation</u>, ed. G. Shaviv and J. Rosen (New York, Wiley), 169.

3. Hill, H. A., and Stebbins, R. T. 1975, Ann. N. Y. Acad. Sci., <u>262</u>, 472.

4. SCLERA is an acronym for the Santa Catalina Laboratory for Experimental Relativity by Astrometry and is a research facility jointly operated by the University of Arizona and Wesleyan University.

5. Cox, J. P. 1974, Rep. Prog. Phys., <u>37</u>, 563.

6. Christensen-Dalsgaard, J., Dilke, F. W. W., and Gough, D. O. 1974, Mon. Not. Roy. Astr. Soc., <u>169</u>, 429.

7. Dilke, F. W. W., and Gough, D. O. 1974, Nature, <u>240</u>, 262.

8. Boury, A., Gabriel, M., Noels, A., Scuflaire, R., and Ledoux, D. 1975, Astr. Ap., <u>41</u>, 279.

9. Unno, W. 1975, Publ. Astron. Soc. Japan, <u>27</u>, 81.

10. Shibahashi, H., Osaki, Y., and Unno, W. 1975, Publ. Astron. Soc. Japan, <u>27</u>, 401.

11. Poor, C. L. 1908, Ann. N. Y. Acad. Sci., <u>18</u>, 385.

12. Musman, S. and Rust, D. M. 1970, Solar Phys., <u>13</u>, 261.

13. Wolff, C. L. 1972, Ap. J. (Letters), <u>177</u>, L87.

14. Ando, H., and Osaki, Y. 1975, Publ. Astron. Soc. Japan, <u>27</u>, 581.

15. Baker, N. H., and Kippenhahn, R. 1962, Z. Astrophys., <u>54</u>, 114.

16. Moore, D. W., and Spiegel, E. A. 1966, Ap. J., <u>143</u>, 871.

17. Wolff, C. L. 1972, Ap. J., <u>176</u>, 833.

18. There are two mechanisms by which the sun may undergo oscillations: the first produces acoustic or pressure waves and the second gravity waves. The physics of acoustic waves within the sun is the same, for example, as for acoustic waves in an organ pipe. The simplest mode is the fundamental which has a period on the order of an hour. The fundamental mode is spherically symmetric: the sun's radius exhibits harmonic motion about some mean value. As is the case with music, this fundamental mode and its spherically symmetric overtones are all radial modes. There are also non-radial modes which lack spherical symmetry; these modes cause distortion of the sun's surface. These non-radial modes are also overtones, the frequencies of which are higher than their spherically symmetric counterparts.

Gravity waves are due to density variations with depth. Just as a fishing bob is pushed up by the water when sunk, so a volume of fluid is pushed up by the buoyant force when immersed in a denser fluid. As the volume

rises above its equilibrium depth, it is pulled back
down by gravity and in this manner, oscillation occurs,
provided the fluids are not to viscous. Gravity waves
by necessity lack spherical symmetry since some of the
fluid must be displaced upwards as the other parts are
displaced downwards. These two modes are often simply
referred to as p and g modes respectively.

19. Deubner, F.-L. 1976, Astr. Ap., 51, 189.

20. Wolff, C. L. 1976, Ap. J., 205, 612.

21. Noyes, R. W. 1967, Proc. IAU Symposium no. 28, 293.

22. Beckers, J. M., and Canfield, R. C. 1975, International
    Colloquium on "Physics of Motions in Stellar Atmo-
    spheres," Nice, France, September 1975 (AFCRL-TR-
    75-0592).

23. Leighton, R. B. 1960, Proc. IAU Symposium no. 12, 321
    (Nuovo Cimento Suppl., 22, 1961).

24. Noyes, R. W., and Leighton, R. B. 1963, Ap. J., 138,
    631.

25. Souffrin, P. 1966, Ann. d'Ap., 29, 55.

26. Stein, R. F. 1967, Solar Phys., 2, 385.

27. Kahn, F. D. 1961, Ap. J., 134, 343.

28. Whitaker, W. A. 1963, Ap. J., 137, 914.

29. Uchida, Y. 1967, Ap. J., 147, 181.

30. Thomas, J., Clark, P. A., and Clark, A., Jr. 1971,
    Solar Phys., 16, 51.

31. Frazier, E. N. 1968, Z. f. Astrophys., 68, 345.

32. Ulrich, R. K. 1970, Ap. J., 162, 193.

33. Leibacher, J. W., and Stein, R. F. 1971, Ap. Lett., 7,
    191.

34. Deubner, F.-L. 1975, Astr. Ap., 44, 371.

35. _____. 1977, IAU No. 36, ed. R. M. Bonnet and P.
    Delache (G. de Bussac).

36. Rhodes, E. J., Jr., Ulrich, R. K., and Simon, G. W. 1977, Ap. J., submitted.

37. Deubner, F.-L. 1972, Solar Phys., 22, 263.

38. Fossat, E., and Ricort, G. 1973, Solar Phys., 28, 311.

39. Fossat, E. 1975, Thesis, Nice University, Nice, France.

40. Livingston, W. C. 1975, in the reference Beckers and Canfield (22).

41. Hill, H. A., Stebbins, R. T., and Brown, T. M. 1976, Atomic Masses and Fundamental Constants, 5, ed. J. H. Sanders and A. H. Wapstra (New York, Plenum), 622.

42. Brown, T. M., Stebbins, R. T., and Hill, H. A. 1976, The Proceedings of the Solar and Stellar Pulsation Conference, Los Alamos, New Mexico, ed. A. N. Cox and R. G. Deupree, Los Alamos Report No. LA-6544-C, 1.

43. Hill, H. A. and Caudell, T. P. 1977, Mon. Not. Roy. Astr. Soc., submitted.

44. Fossat, E., and Ricort, G. 1975, Astr. Ap., 43, 243;253.

45. Grec, G., and Fossat, E. 1977, Astr. Ap., 55, 411.

46. Brookes, J. R., Isaak, G. R., and van der Raay, H. B. 1976, Nature, 259, 92; 1977, New Scientist, 73, 294.

47. Severny, A. B., Kotov, V. A., and Tsap, T. T. 1976, Nature, 259, 87.

48. Livingston, W. C., Milkey, R., and Slaughter, C. 1977, Ap. J., 211, 281.

49. Dittmer, P. H., Scherrer, P., and Wilcox, J. M. 1977, Book of Abstracts, Topical Conference on Solar and Interplanetary Physics, January 12-15, 1977, Tucson Arizona, 16.

50. Musman, S. and Nye, A. H. 1977, Ap. J. (Letters) 212, L95.

51. Beckers, J. M. and Ayres, T. R. 1977, Ap. J. (Letters) submitted.

52. Brown, T. M., Stebbins, R. T., and Hill, H. A. 1977, Ap. J., submitted.

53. Groth, E. J. 1975, Ap. J. Suppl, 29, 285.

54. Hill, H. A., and Stebbins, R. T. 1975, Ap. J., 200, 471.

55. Christensen-Dalsgaard, J., and Gough, D. O. 1976, Nature, 259, 89.

56. Rouse, C. A. 1977, Astr. Ap., 55, 477.

57. Scuflaire, R., Gabriel, M., Noels, A., and Boury, A. 1975, Astr. Ap., 45, 15.

58. Iben, I., Jr. 1976, Ap. J. (Letters), 204, L147.

59. Iben, I., Jr., and Mahaffy, J. 1976, Ap. J. (Letters), 209, 39.

60. Wolff, C. L. 1977, private communication.

61. Kotov, V. A., Severny, A. B., and Tsap, T. T. 1977, Mon. Not. Roy. Astr. Soc., submitted.

62. Worden, S. P., and Simon, G. W. 1976, Ap. J. (Letters), 210, L1.

63. Dicke, R. H., and Goldenberg, H. M. 1974, Ap. J. Suppl., 27, 131.

64. Dicke, R. H. 1976, Solar Phys., 37, 271.

65. _____. 1976, Phys. Rev. Lett., 37, 1240.

66. Gough, D. O. 1977, The Solar Output and its Variations, ed. O. R. White (University of Colorado Press, Boulder, Colorado).

67. _____. 1977, Mon. Not. Roy. Astr. Soc., submitted.

68. Deubner, F.-L. 1977, Astr. Ap., 57, 317.

69. Oleson, J. R., Zanoni, C. A., Hill, H. A., Healy, A. W., Clayton, P. D., Patz, D. L. 1974, Appl. Opt., 13, 206.

70. Hill, H. A., Stebbins, R. T., and Oleson, J. R. 1975, Ap. J., <u>200</u>, 484.

71. Hill, H. A., Caudell, T. P., and Rosenwald, R. D. 1976, The Proceedings of the Solar and Stellar Pulsation Conference, Los Alamos, New Mexico, ed. A. N. Cox and R. G. Deupree, Los Alamos Report No. LA-6544-C, 7.

72. Hill, H. A., Rosenwald, R. D., and Caudell, T. P. 1977, Ap. J., submitted.

73. Leighton, R. B., Noyes, R. W., and Simon, G. W. 1962, Ap. J., <u>135</u>, 474.

74. Tanenbaum, A. S., Wilcox, J. M., Frazier, E. N., and Howard, R. 1969, Solar Phys., <u>9</u>, 328.

75. Hill, H. A., Caudell, T. P., and Rosenwald, R. D. 1977, Ap. J. (Letters), <u>213</u>, L81.

76. Hill, H. A., Livingston, W. C., and Caudell, T. P. 1977, Ap. J. (Letters), <u>214</u>, L137.

77. Edmonds, A. R. 1957, <u>Angular Momentum in Quantum Mechanics</u> (Princeton University Press, Princeton, New Jersey).

78. Dittmer, P. H. 1977, Ph.D. Thesis, Stanford University, Stanford, California.

79. Schatzman, E. 1956, Ann. d'Ap., <u>19</u>, 45.

80. Unno, W. 1965, Publ. Astron. Soc. Japan, <u>17</u>, 205.

81. Zhugshda, Y. D. 1972, Solar Phys., <u>25</u>, 329.

82. Stein, R. F., and Leibacher, J. 1974, Annual Review of Astronomy and Astrophysics, Vol. 12, ed. G. Burbridge, D. Layzer, and J. Philips, 407.

83. Ulrich, R. K., and Rhodes, E. J., Jr. 1977, Ap. J., submitted.

84. Unno, W., and Spiegel, E. A. 1966, Publ. Astron. Soc. Japan, <u>18</u>, 85.

85. On the low temperature side of the hydrogen ionization zone which occurs at $T \sim 10^4$ K, the opacity, κ, increases rapidly with temperature in contrast to its behavior in

most other regions of the sun. For example, upon compression of the gas in this particular region by a wave and
its accompanying increase in temperature, there is a
local increase in κ which diminishes the flow of radiation. However, this reduced flow of radiation enhances
the local rate of temperature increase produced by the
wave, further contributing to an increase in κ. This
positive "feedback" may produce an anomalous local
effect or can actually lead to an instability of a star
against pulsation. This direct effect of opacity variations has been called the κ-mechanism. For further
discussion, see (5).

86. Holweger, H., and Testerman, L. 1975, Solar Phys., 43,
       271.

87. Spiegel, E. A. 1957, Ap. J., 126, 202.

88. Gough, D. O. 1977, IAU Colloq. No. 36, ed. R. M.
       Bonnet and P. Delache (G. de Bussac).

89. Frazier, E. N. 1968, Ap. J., 152, 557.

90. Evans, J. W., and Michard, R. 1962, Ap. J., 135, 812.

91. _____. 1962, Ap. J., 136, 487.

92. Edmonds, F. N., Jr., Michard, R., and Servajean, R.
       1965, Ann. Astrophys., 28, 534.

93. Roddier, F. 1966, Ann. Astrophys., 29, 639.

94. Howard, R. 1967, Solar Phys., 2, 3.

95. Deubner, F.-L. 1967, Solar Phys., 2, 133.

96. Livingston, W. C. 1968, Ap. J., 153, 929.

97. Cox, A. N., and Tabor, J. E. 1976, Ap. J. Suppl., 31,
       271.

98. Parnell, R. L., and Beckers, J. M. 1969, Solar Phys.,
       9, 35.

99. Beckers, J. M., and Milkey, R. W. 1975, Solar Phys.,
       43, 289.

100. Howard, R., and Harvey, J. 1970, Solar Phys., 12, 23.

101. Gilman, P. A. 1974, Ann. Rev. Astron. & Astrophys., 12,
       47.

102. Ledoux, P., and Walraven, Th. 1958, Handbuch der Physik, Vol. LI., ed. S. Flugge (Berlin, Springer-Verlag), 353.

103. Ledoux, P. 1951, Ap. J., <u>114</u>, 373.

104. Deubner, F.-L. 1977, Bull. American Astr. Soc., <u>9</u>, 375.

105. Rhodes, E. J., Jr., and Ulrich, R. K. 1977, Bull. American Astr. Soc., <u>9</u>, 336.

106. Christensen-Dalsgaard, J., and Gough, D. O. 1975, Mémoires Société Royale des Sciences de Liège, 6e série, tome VIII, 309.

107. Hill, H. A., Rosenwald, R. D., Ballard, P. T., and Bryan, S. P. 1977, SIAM Journal on Mathematical Analysis, submitted.

108. Uberoi, M. S. 1955, Ap. J., <u>122</u>, 466.

109. Gradshteyn, I. S., and Ryzhik, I. M. 1965, <u>Tables of Integrals, Series, and Products</u> (New York, Academic Press).